Information and Instructions

This shop manual contains several sections each covering a specific group of wheel type tractors. The Tab Index on the preceding page can be used to locate the section pertaining to each group of tractors. Each section contains the necessary specifications and the brief but terse procedural data needed by a mechanic when repairing a tractor on which he has had no previous actual experience.

Within each section, the material is arranged in a systematic order beginning with an index which is followed immediately by a Table of Condensed Service Specifications. These specifications include dimensions, fits, clearances and timing instructions. Next in order of arrangement is the procedures paragraphs.

In the procedures paragraphs, the order of presentation starts with the front axle system and steering and proceeding toward the rear axle. The last paragraphs are devoted to the power take-off and power lift systems. Interspersed where needed are additional tabular specifications pertaining to wear limits, torquing, etc.

HOW TO USE THE INDEX

Suppose you want to know the procedure for R&R (remove and reinstall) of the engine camshaft. Your first step is to look in the index under the main heading of ENGINE until you find the entry "Camshaft." Now read to the right where under the column covering the tractor you are repairing, you will find a number which indicates the beginning paragraph pertaining to the camshaft. To locate this wanted paragraph in the manual, turn the pages until the running index appearing on the top outside corner of each page contains the number you are seeking. In this paragraph you will find the information concerning the removal of the camshaft.

More information available at Clymer.com
Phone: 805-498-6703

Haynes Publishing Group
Sparkford Nr Yeovil
Somerset BA22 7JJ England

Haynes North America, Inc
859 Lawrence Drive
Newbury Park
California 91320 USA

SHOP MANUAL

JOHN DEERE

MODELS 850, 950 and 1050

Tractor serial number is located on plate attached to rear machined face of tractor center housing directly below pto output shaft. Engine serial number is stamped in machined pad on left side of engine near front.

INDEX (By Starting Paragraph)

CONDENSED SERVICE DATA

GENERAL	850	950	1050
Engine Make	Yanmar	Yanmar	Yanmar
Engine Model	3T80J	3T90J	3T90TJ
Number of Cylinders	3	3	3
Bore-mm	80	90	90
Stroke-mm	85	90	90
Displacement-cm^3	1281	1717	1717
Compression Ratio	21:1	20:1	21:1
TUNE-UP			
Firing Order	1-3-2	1-3-2	1-3-2
Valve Tappet Gap-Cold			
Exhaust-mm	0.20	0.15	0.15
Inlet-mm	0.20	0.15	0.15
Governed Speeds-Engine RPM			
Low Idle	850	850	850
High Idle	2750	2600	2600
Loaded	2600	2400	2400
Horsepower at pto*	22.27	27.36	33.41
Battery –			
Volts	12	12	12
Ground Polarity	Negative	Negative	Negative
*According to Nebraska Test at Loaded RPM.			
CAPACITIES			
Cooling Systems			
(Liters)	5.5	6	6.7
Crankcase (Liters)**	4.5	6.4	6.4
Front Drive Axle (Liters)		8.5	8.5
Fuel Tank (Liters)	32	32	42
Transmission &			
Hydraulic (Liters)	18	18	26

**With Filter.

FRONT AXLE AND STEERING GEAR

FRONT AXLE ASSEMBLY

Two Wheel Drive Models

1. **ADJUSTMENT.** Refer to Fig. 1 for exploded view of adjustable tread front axle and to Fig. 3 for exploded view of axle extensions.

Toe-in should be 3-9 mm. Length of drag link should be adjusted to provide equally sharp turns in both directions. Refer to the following recommended length of drag link.

850

Tread Width	Drag Link Length
1070 mm (42 in.)	790 mm (31 in.)
1170 mm (46 in.)	793 mm (31¼ in.)
1270 mm (48 in.)	800 mm (31½ in.)

950

Tread Width	Drag Link Length
1140 mm (45 in.)	853 mm (33½ in.)
1240 mm (49 in.)	859 mm (33¾ in.)
1340 mm (53 in.)	865 mm (34 in.)
1440 mm (57 in.)	875 mm (34½ in.)

Front wheel bearings should be removed, cleaned, inspected, and renewed if damaged or repacked with new grease after each 600 hours of operation. Wheel lug bolts should be tightened to 118-147 N·m torque.

2. **REMOVE AND REINSTALL.** Support front of tractor and disconnect drag link from steering arm. Remove

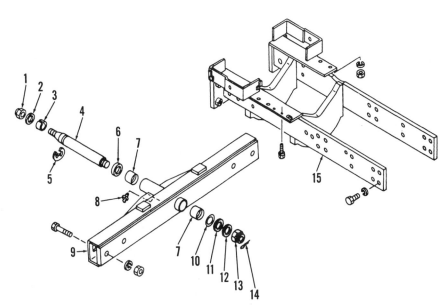

Fig. 1 – Exploded view of axle main member and support frame used on two wheel drive models.

1. Nut	5. Snap ring	9. Axle main member
2. Washer	6. Seal	10. Seal
3. Spacer	7. Bushing	11. Washer	13. Castle nut
4. Pivot pin	8. Grease fitting	12. Shim washer	14. Cotter pin
			15. Support frame

nut (1 – Fig. 1) and snap ring (5) from front of pivot shaft (4); then remove cotter pin (14), castle nut (13) and washer from rear of pivot pin. Support axle with jack under center then remove pivot pin using a nut, washers and spacer to pull pivot pin from bracket. The spacer can be locally manufactured from 38 mm ID pipe which is 19 mm long. Raise tractor, then carefully move axle forward from under tractor.

Diameter of pivot pin should be 34.95-34.98 mm at bearing surfaces. Inside diameter of bearings (7) should be 35.0-35.04 mm. Pin to bushing clearance should be 0.03-0.10 mm but should not be more than 0.41 mm. New bushings should be installed flush with ends of tube.

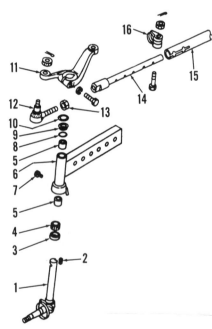

Fig. 2 – Use feeler gage as shown to measure end play of axle main member.

Position axle between supports, then install pivot pin from the front. Install thrust washer (2), tighten nut (1) to 323-402 N·m torque, then install snap ring (5). Install seal (10), washers (11 & 12) and nut (13) on rear of pivot pin. Tighten nut (13) securely, but do not install cotter pin (14). Pull axle fully forward on pivot pin, then measure end play using feeler gage between rear of axle and axle housing rear support as shown in Fig. 2. End play should be 0-0.13 mm. If end play is excessive, tighten pivot pin rear nut and recheck clearance with feeler gage. Be sure that axle is free to pivot after tightening rear nut. Install cotter pin when nut is correctly positioned, then lubricate pivot pin with multi-purpose grease. If tightening nut does not decrease end play sufficiently, install additional shim washer (12 – Fig. 1) and recheck.

3. **OVERHAUL.** The steering knee (Fig. 3) is equipped with renewable bushings (5), thrust bearing (4) and seals (3 & 8). Inside diameter of bushings (5) should be 30.0-30.03 mm. Spindle (1) should not be scored or worn. Clearance between spindle (1) and bushings (5) should be 0.03-0.08 mm with wear limit of 0.25 mm. New bushings should be flush with ends of axle tubes. End play of spindle (1) should be 0.03-0.61 mm after assembly is complete. If necessary,

install shims (10) to reduce end play. Tighten steering arm clamp screw to 79-98 N·m torque. Screws clamping extension (6) in axle (9 – Fig. 1) should be tightened to 167-206 N·m torque.

Four Wheel Drive Models

4. **ADJUSTMENT.** Refer to appropriate paragraphs 25 through 30 for service to individual units. Toe-in should be 3-9 mm. Length of drag link should be adjusted to provide equally sharp turns in both directions. Recommended drag link length should be 830 mm (32 inches).

5. **REMOVE AND REINSTALL.** To remove the front axle from four wheel drive models, first block rear wheels. Disconnect headlight wire, disconnect hood holding bracket, then unbolt and remove hood. Disconnect battery cables, then remove battery. Remove locknuts (N – Fig. 4) from the screws that attach front axle to support. Loosen drive shaft front cover clamp, remove attaching screws, then slide cover toward rear. Disconnect spring loaded collar from front drive pinion. Move drive shaft and cover assembly out of the way. Be careful not to lose steel balls from front and rear of drive shaft collars. Detach drag link from steering arm. Support weight of tractor by attaching overhead hoist to front weight support and place

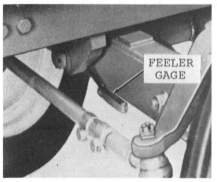

Fig. 3 – Axle extension used on two wheel drive models.

1. Spindle	9. Retainer
2. Key	10. Shim
3. Seal	11. Steering arm
4. Thrust bearing	12. Tie rod end
5. Bushings	13. Nut
6. Axle extension	14. Tie rod
7. Bolt	15. Tube
8. Seal	16. Clamp

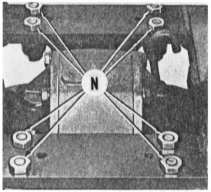

Fig. 4 – View of nuts (N) used to lock screws that attach front wheel drive axle to support frame.

Fig. 5 – Support tractor with chain hoist and lower axle with floor jack.

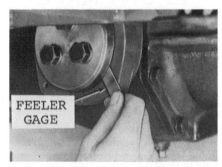

Fig. 6 – Measure end play with feeler gage as described in text.

Fig. 7 – Shims shown are installed to adjust axle pivot end play. Refer to text.

jack under center of front axle. Raise tractor with floor jack and remove wheels. Remove all eight screws which attach axle mounting bracket to front support. Lower axle away from tractor.

When installing, reverse the removal procedure. Be sure that dowel pins in axle mounting brackets are correctly engaging holes in support. Coat threads of attaching screws with "Loctite" or equivalent and tighten to 149 N·m. Remove cap from axle front support, drive axle forward, then measure end play with feeler gage as shown in Fig. 6. End play should be 0.05-0.30 mm. Remove plate and add or deduct shims (Fig. 7) as required if end play is incorrect. Tighten plate retaining cap screws to 147 N·m torque.

STEERING GEAR

All Models

10. **REMOVE AND REINSTALL.** Pry cap from center of steering wheel, remove steering wheel retaining nut, then use puller to remove steering wheel from shaft. Disconnect battery ground cable and remove access panel from rear of console. Disconnect drive cable from tachometer, fuel lines from tank and throttle linkage from lever on right side. Remove two screws from bottom of each side of console and two screws from bottom of front, then lift console up off steering column.

The steering gear can be unbolted and removed from clutch housing after disconnecting drag link from steering arm; however, most service can be accomplished while leaving steering gear attached to clutch housing.

Reinstall steering gear by reversing the removal procedure. If removed from clutch housing, coat threads of attaching screws with sealer and tighten to 98-117 N·m torque.

11. **OVERHAUL.** Remove cap screws attaching side cover (16 – Fig. 10) to housing (21), remove locknut (14), then turn adjusting screw (15) in to push cover away from housing. Loosen nut (25) and turn shaft (6) until gear on shaft (19) is aligned with opening in housing. Remove adjusting screw (15), then bump end of shaft (19) out toward right side of tractor. After steering arm (24) is released from taper on end of shaft. remove nut (25) and steering arm withdraw shaft (19) from housing Unbolt steering column (12 withdraw shaft (6) and ball nut (8). Steering column is screwed into housing and locked in place with a large locknut on some models.

Diameter of bushings (22) should be 28.60-28.62 mm for some models, 31.75-31.77 mm for other models. Bushings are available only as an assembly with housing (21). Bearing journals should be 28.57-28.59 mm or 31.716-31.741 mm diameter. Seal (23) should be installed flush with housing.

Inner race for ball bearings (7) is integral with shaft (6). Ball nut (8) is available only with shaft (6) and disassembly is not recommended. Clean ball nut and check for smoothness. Apply liberal coating of grease to shaft and ball nut. Grease lower bearing (7) and

position in bearing race located in housing. Grease upper bearing (7) and locate on shaft (6). Position shaft (6) and ball nut in housing (21) then lower shim gasket (13) and column (12) over shaft (6). Turn shaft while pressing down to seat bearings in races. End play of shaft (6) should be limited to 0.025-0.102 mm after column (12) is installed. Shims (13) are used to set end play on models which have column (12) attached with four screws. End play for models with threaded column is set by turning column into housing until correct end play is achieved, then locking position with large locknut.

Turn shaft (6) until ball nut (8) is in center of travel. Coat cross shaft (19) with grease, then install with center tooth in center valley of ball nut. Position adjustment screw (15) and shim (18) in end of shaft and gasket (20) on housing, then install cover (16). Turn adjuster screw (15) out through cover (16) while installing, then tighten the four attaching screws to 23-30 N·m torque. Be sure that adjusting screw (15) remains loose while tightening the four attaching screws. Install steering arm (24) over shaft splines with index marks on shaft and splines aligned. On some models, a missing spline on shaft should be aligned with mark on steering arm. Tighten nut (25) to 147-196 N·m on models with small (28.57-28.59 mm) journal cross shaft; 205-245 N·m for models with larger (31.716-31.741 mm) journal cross shaft. Adjust screw (15) to provide 25-50 mm free play at rim of steering wheel, then lock adjustment with nut (14). Initial adjustment can be accomplished by turning screw clockwise to reduce end play to zero then back screw up ¼-turn and lock with nut (14). Adjustment screw (15) is accessible through hole in right side of console. Fill steering gear with 30 mL of John Deere Multi-Purpose Lubricant.

POWER STEERING

A flow divider valve is used on models with power steering to assure a constant volume of oil to the power steering system. The control valve is located on the drag link and pressurizes an assist cylinder located between spindle steering arm and bracket attached to clutch housing.

FLOW DIVIDER BLOCK

All Models So Equipped

15. **ADJUSTMENT.** To check and adjust relief pressure, proceed as follows: Install "T" fitting between one of the

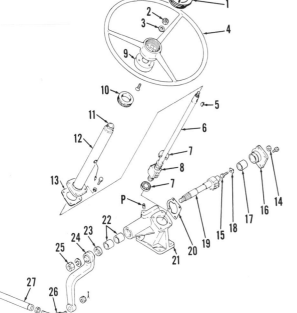

Fig. 10 – Exploded view of steering gear assembly. Column (12) is threaded into housing (21) of some models.

P. Plug
1. Cap
2. Nut
3. Washer
4. Steering wheel
5. Woodruff key
6. Steering shaft
7. Bearings
8. Ball nut
9. Collar
10. Grommet
11. Insert
12. Column
13. Shim gasket
14. Locknut
15. Adjusting screw
16. Side cover
17. Bushing
18. Shim
19. Sector gear and cross shaft
20. Gasket
21. Housing
22. Bushings
23. Seal
24. Steering arm
25. Nut
26. Drag link end
27. Drag link

cylinder ports and steering lines, then connect a 0-350 bar pressure gage as shown in Fig. 12 or 13. Run engine at 1500 rpm, turn steering wheel to extreme left or right as necessary to pressurize gage, hold steering wheel in this position and observe pressure. Recommended pressure should be 127.6-134.5 bar (1850-1950 psi). To adjust, remove cap from relief valve adjusting screw

Fig. 12 — View of pressure gage attached to junction block for checking steering relief pressure on 850 and 950 models.

A. Junction block
B. Adapter for plug
   opening
C. Hose
D. Gage
V. Valve

Fig. 13 — View of pressure gage attached to steering control valve to check steering relief pressure on 1050 models.

A. Control valve
B. Steering cylinder
C. Hose
D. Gage

Fig. 14 — View of diverter valve assembly (1) on 850 and 950 models showing location of relief valve adjustment screw (3) and locknut (2).

(3 – Fig. 14 or Fig. 15), loosen locknut (2) then adjust by turning screw (3). Turning screw clockwise increases pressure.

Individual parts of relief valve are not available separately and complete valve must be renewed if valve is damaged. Be sure that main system is not faulty if sufficient steering pressure can not be attained. Also, check condition of flow control valve (4 – Fig. 16). Flow should be within limits of 6.1-6.8 liters/min. (1.6-1.8 gpm) at 104 bar (1500 psi) with engine at 1500 rpm.

CONTROL VALVE

All Models So Equipped

18. **R&R AND OVERHAUL.** Be sure to measure distance from center of front ball joint (rod end) to center of rear ball joint (length "A" – Fig. 18) before removing from tractor. Record this measurement to facilitate assembly. Clean control valve and surrounding area thoroughly. Disconnect hoses from control valve, then immediately cap all

Fig. 15 — View of diverter valve assembly (1) on 1050 models showing location of relief valve adjustment screw (3) and locknut (2).

valve ports and hoses to prevent contamination. Detach rear rod end from steering pitman arm. Loosen locknut (2) at front of valve and loosen clamp (3) at rear of valve. Remove parts (1, 2 & 4) from rear of control valve. Thread steering valve from tube (14), counting turns for removal to assist installation adjustment.

Remove retainer (10), drive roll pin (11) out, then withdraw sleeve (5). Spool and parts (6 through 8) can be withdrawn from valve housing. Spring (8) should exert 178 N when compressed to 12.7 mm. Examine spool valve and housing for any evidence of nicks, burrs or scoring. Small nicks or imperfections can be removed using fine crocus cloth.

Coat all parts with clean John Deere Hy-Guard Transmission and Hydraulic oil prior to assembly. Be sure to keep all parts and working area clean while assembling. Always install new "O" rings and back-up rings when assembling.

Install "O" ring (9) in groove at front of housing, then install back-up ring (13). Install "O" ring (9) in front of housing, then carefully insert assembled valve spool with parts (6 through 8) into housing bore while turning slightly to prevent sticking. Insert sleeve (5) over spool and install pin (11). Install retainer ring (10) in groove around outside of housing.

When reinstalling, thread control valve onto rod (14) the same number of turns as it was removed, then tighten locknut (N) against rod. Install rod end (1) into rod (4) and adjust length "A" to the same dimension that it was before it was removed. Ports in control valve housing should be toward top.

To check length adjustment, disconnect rod end from steering pitman arm, turn steering wheel to full right stop, then adjust length of control valve rod so that rod end just fits into steering pit-

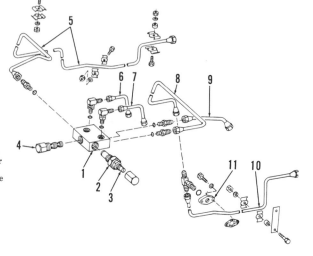

Fig. 16 — Power steering lines should be routed as shown from the diverter valve (1). Shape of pipes may be slightly different on some models.

1. Diverter valve block
2. Relief valve locknut
3. Adjusting screw
4. Flow control valve
5. Pressure pipe to power steering
6. Return oil pipe
7. Pipe to selective control valve
8. Upper return pipe from power steering
9. Pressure pipe to diverter valve from control block
10. Power steering return pipe
11. Cover

man arm. Shorten rod two full turns more, then attach to steering pitman arm. Make sure that all locknuts and clamps are fully tightened. Nuts on rod ends should be tightened to 75-81 N·m torque and locked with cotter pin.

STEERING CYLINDER

All Models So Equipped

20. **R&R AND OVERHAUL.** Be sure to measure distance from center of front ball joint (rod end) to center of rear ball joint (length "B" – Fig. 18) before removing from tractor. Record this dimension to facilitate assembly. Clean cylinder and surrounding area thoroughly. Disconnect hoses from control valve, then immediately cap both ports and both hoses to prevent contamination. Detach control valve front rod end from link (28), then detach both rod ends of steering cylinder from bracket and axle steering arm.

To disassemble, remove rear rod end from piston rod (21). Remove lock ring (26) through notch in cylinder. Wire can not be reused after removal. Pull rod guide (18), rod (21), piston (24) and associated parts from cylinder. Use two nuts locked together on outside end of rod (21) to hold rod while removing nut (25) and piston (24).

Piston diameter is 34.72-34.79 mm diameter and cylinder (27) is 34.93-35.00 mm. Examine piston and cylinder carefully for nicks, wear or scoring. Slight nicks or imperfections can be removed using fine crocus cloth.

Coat all parts with clean John Deere Hy-Guard Transmission and Hydraulic oil prior to assembly. Be sure to keep all parts and working area clean while assembling. Always install new "O" rings and back-up rings when assembling.

Install "O" ring (17) in groove of rod guide, then install back-up washer (16) toward outside. Install "O" ring (20) and back-up ring (19) in outside groove of rod guide. Install wiper seal (15) in rod guide flush with bore and with lips toward outside. Position rod guide (18) over piston rod (21), then install piston (24). Tighten nut (25) to 68-81 N·m. Install "O" ring (23) in groove of piston, then position slipper ring (22) over "O" ring. Carefully slide piston into cylinder, then rod guide into cylinder. Install new lock ring (26) to hold rod guide in cylinder. Insert lock ring through slot in cylinder and into hole in rod guide. Rotate rod guide until lock ring is pulled into place.

When installing, link (28) should be threaded fully into rod (29), if used, and fully into cylinder (27). Link (28) must be aligned with hole vertical so that control

valve rod end can be attached and ports in cylinder (27) and rod guide (18) must be down. Tighten clamp and locknut to maintain correct alignment. Attach rod end to rod (21), turn steering wheel to full right and front wheel to full right. Push piston rod (21) fully into cylinder and adjust rod ends to align with hole in pitman steering arm. Turn rod end three more turns onto rod, then tighten clamp and attach to pitman steering arm. Tighten all rod end nuts to 75-81 N·m and install cotter pin.

FOUR WHEEL DRIVE

OUTER DRIVE HOUSING

All Models So Equipped

25. **R&R AND OVERHAUL.** The complete outer drive housing can be separated from axle after removing screws (A – Fig. 25) or disassembly can

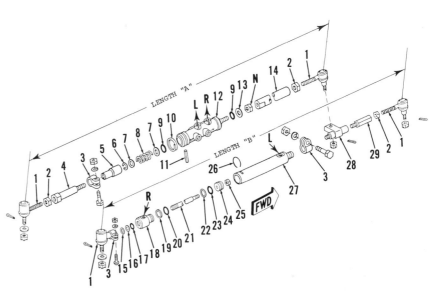

Fig. 18 – Exploded view of power steering control valve and cylinder. Individual parts of control valve (12) are not available separately. For left turn, ports (L) are pressurized; for right turn, ports (R) are pressurized.

1. Rod ends (not all alike)	9. "O" rings (2 used)	16. Back-up washer	23. "O" ring
2. Locknuts	10. Retaining ring	17. "O" ring	24. Piston
3. Clamps	11. Dowel pin	18. Rod guide	25. Locknut
4. Rod	12. Control valve	19. Back-up ring	26. Lock ring
5. Sleeve	13. Back-up washer	20. "O" ring	27. Cylinder
6. Snap ring	14. Rod	21. Piston rod	28. Link
7. Washers (2 used)	15. Wiper seal	22. Slipper ring	29. Rod (not used on 850)
8. Spring			

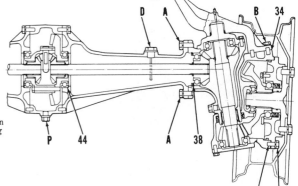

Fig. 25 – Cross-section of front drive axle. Thickness of shims (7, 34, 38 & 44) determines bearing and gear adjustment. Parts are also shown in Figs. 27, 28 & 30.

A. Screws attaching outer drive housing
B. Screws attaching steering arm
C. Screws attaching axle housing cover
D. Oil level dipstick
P. Drain plug

be accomplished before detaching outer end of axle from center section.

To remove the complete outer drive housing, remove wheel, detach drag link and/or tie rod from steering arm and support axle housing to prevent tipping. Support outer drive housing separately, then remove the six screws (A). Move outer drive housing off dowel pins away from axle center section. Disassembly of outer drive assembly can be accomplished as outlined in following paragraphs.

Disassembly can be accomplished with outer drive housing attached to or separated from axle center section. If attached, remove wheel and support axle from tipping. Remove eight screws (C–Fig. 26) then use two of the removed screws in threaded holes (J) for pushing cover away from housing. Be sure to record and save the shims (7–Fig. 27) located between cover and housing.

Bend tang washer (13) out of notch in spanner nut (14), then remove nut and tang washer. Bump axle (1) out of bearing (12) and gear (11). Seal wear sleeve (3) will remain on axle (1). Unbolt bearing retainer plate (10), then use a soft drift to drive gear and bearing (8) from cover (4).

Support outer drive housing, then remove four screws (B–Fig. 26). Separate outer drive housing (15–Fig. 28) from spindle housing (25). Be careful to save and record thickness of shims (34) located between steering arm (35) and outer drive housing (15). Steering arm (35) can be lifted from axle.

Unbolt spindle upper cap (31), then bump spindle shaft (26), spindle cap (31), gear (28) and bearing (29), **UP** out of housing (25). Bevel gear (36), bearing (37), thrust washer (39) and shims (38) can be removed after removing snap ring (40).

If not already removed, remove screws (A) and lift end housing from axle center section.

Clean and check all parts, especially seals, bearings and journals for roller bearings. Renew all seals and "O" rings upon reassembly. Inside diameter of bushing (32) should be 35.02-35.07 mm.

Thickness of shims (7 & 38 – Fig. 25) is used to adjust backlash between gears; shims (34) are used to adjust spindle end play. Refer to the following for selection of shims and reassembly.

Install bearing (24 – Fig. 28) and snap ring (23) in lower bore of housing (25).

Assemble spindle (26), snap ring (27), gear (28) and bearing (29), then, locate in spindle housing (25). Install gasket (29) and cap (31). Tighten cap retaining screws to 23-30 N·m torque.

Install gear (36) and bearing (37) in bore of housing (25). Bump bearing into bore as far as possible, then install thrust washer (39) and snap ring (40). Measure gap between bearing and thrust washer with feeler gage, then install shims (38) necessary to reduce measured clearance to 0.07-0.17 mm. Remove snap ring (40) and thrust washer (39) to install shims, then

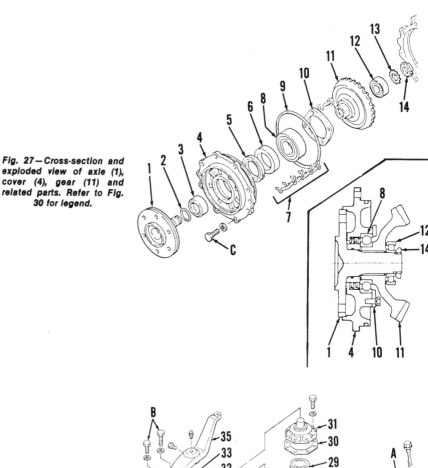

Fig. 27 — Cross-section and exploded view of axle (1), cover (4), gear (11) and related parts. Refer to Fig. 30 for legend.

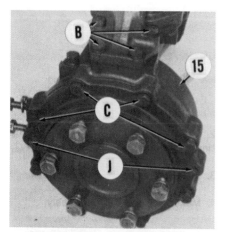

Fig. 26 — Eight screws (C) attach cover to housing. Threaded holes (J) can be used with jack screws to push cover away from housing (15). Screws (B) attach steering arm to housing.

Fig. 28 — Exploded view of spindle (26), housing (25), gear housing (15) and related parts. Refer to Fig. 30 for legend.

reinstall thrust washer and snap ring. Measure clearance again after shims are installed. **Be sure that open section of snap ring is aligned with notch at bottom of housing to assure proper lubrication.**

Install thin outer seal (5 – Fig. 27) with open side facing outward (axle flange). Install inner seal (6) with open side facing inward (gear). "O" ring (2) in wear sleeve groove should hold wear sleeve (3) onto axle journal. Assemble bearing (8), gear (11) and retainer plate (10) to cover (4). Install inner race of bearing (12) onto axle with flange against gear (11), then install bearing outer race and rollers. Position tang washer (13) on axle with inner tab in notch, then install nut (14). Assemble gear (18 – Fig. 28) and housing (15) as described in following paragraph, then select shims (7 – Fig. 27) to provide correct backlash before installing "O" ring (9) and completing installation.

Install wiper seal (16 – Fig. 28) in housing bore with open side of seal toward top. Install oil seal (17) in bore with spring loaded lip and open side of seal down. Install lower bevel gear (18), bearing (19), gasket (20) and cover (21). Work through upper opening and bump gear (18) down to be sure that gear and bearing are seated. Install axle and gear assembly (Fig. 27) without "O" ring (9), using shims (7) that were originally installed. **Shims (7) should be evenly distributed in four groups.** Measure gear backlash by using a rocker type dial indicator on teeth of gear (18 – Fig. 28) through hole in housing (15). Backlash should be 0.10-0.15 mm. Change backlash by varying thickness of shims (7 – Fig. 27), but be sure that each of the four groups are exactly the same thickness. Recheck backlash if shims are changed. After selecting correct thickness of shims, remove axle and cover assembly (Fig. 27), install sealing "O" ring (9), then reinstall using selected shims (7). Tighten screws (C) to 55 N·m torque.

Lubricate lower end of spindle housing (25 – Fig. 28), then install axle and housing assembly onto spindle assembly. Lubricate bushing journal on cap (31), then install arm (35) using shims (34) that were removed. Measure end clearance using a feeler gage between arm (35) and cap (31). End play should be 0.02-0.14 mm and can be adjusted by varying thickness of shims (34). Be sure that shims are evenly divided between the two stacks for each arm. Tighten screws (B) to 64-83 N·m torque.

Attach assembled outer drive housing to end of axle using new "O" ring (41).

DIFFERENTIAL AND BEVEL GEARS

All Models So Equipped

30. **R&R AND OVERHAUL.** Remove axle assembly from tractor as outlined in paragraph 5. Unbolt left axle housing (42 – Fig. 31) from center housing (57), then remove differential. Right axle housing (99) can be unbolted and separated from center housing. Unbolt and remove pinion and carrier (62) from center housing.

NOTE: Shims (58) adjust pinion mesh position; shims (68) together with position of nut (69) adjust preload of bearings (60 &

63). **Shims (44 & 100) adjust carrier bearings (45). Moving shims (44 & 100) from one side to the other moves differential and therefore adjusts bevel gear backlash. Be careful not to damage or lose shims when disassembling.**

To assemble pinion, press outer races of bearing (60 & 63) into pinion housing (62). Make sure that both outer races are seated in bottom of bores, then press spacer (64) against outer race of bearing (63). Press inner race and bearing cone of front bearing (60) onto pinion shaft (59) then insert into housing (62). Install inner race and cone of rear bearing (63), then install "O" ring (65) in groove of shaft. Press oil seal (67) into bore of

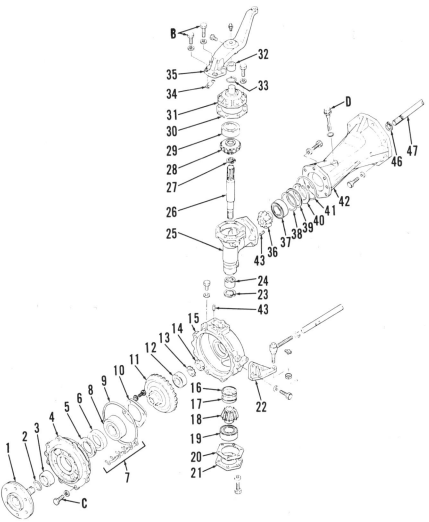

Fig. 30 — Exploded view of left axle housing (42), spindle housing (25), gear housing (15) and related parts.

A. Screws attaching outer drive housing	7. Shims	20. Gasket	33. "O" ring
B. Screws attaching steering arm	8. Ball bearing	21. Cover	34. Shims
C. Screws attaching axle housing cover	9. "O" ring	22. Tie rod arm	35. Steering arm
D. Oil level dipstick	10. Bearing retainer	23. Snap ring	36. Drive gear
1. Axle and hub	11. Gear	24. Roller bearing	37. Ball bearing
2. "O" ring	12. Roller bearing	25. Spindle housing	38. Shims
3. Wear sleeve	13. Tang washer	26. Spindle	39. Thrust plate
4. Cover	14. Spanner nut	27. Snap ring	40. Snap ring
5. Thin seal	15. Gear housing	28. Top gear	41. "O" ring
6. Seal	16. Seal	29. Ball bearing	42. Axle housing
	17. Seal	30. Gasket	43. Roll pin
	18. Lower gear	31. Cap	46. Drive gear
	19. Ball bearing	32. Bushing	47. Drive axle

housing with open side of seal against spacer (64). Lubricate wear sleeve (66) and press onto shaft over "O" ring (65). Install previously removed shims (68) and nut (69). Clamp housing (62) in vise and use torque wrench on nut (69) to measure starting torque necessary to rotate pinion shaft in bearings. Correct starting torque is 0.5-2.0 N·m (3-14 in.-lbs.). Slot in nut must also align with hole in shaft when correct starting torque is obtained. A combination of nut tightening torque and thickness of shims (68) will permit installation of cotter key and correct starting torque.

When assembling attach pinion and housing to center housing using shims (58) that were removed upon disassembly. Install left side carrier bearing (45) in left axle housing (42), using original shims (44), then position differential and ring gear (91 through 98) in bearing with ring gear teeth up. Position gasket (49) on axle housing, then attach center housing to left axle housing.

Be sure that differential is seated in bearing, then measure backlash between ring gear and pinion at three locations around ring gear. Large deviations in recorded backlash indicates improper seating of differential and bearing. Correct backlash is 0.18-0.23 mm when measured with dial indicator and checked at center of ring gear tooth face. Add shims (44) to reduce backlash, remove thickness from shims (44) to increase backlash.

Pinion depth or mesh position can be checked as shown in Fig. 33 using special tool (JDG-48-2) available from manufacturer. Gage is Go/No-Go type and should be used between end of pinion (59) and special machined section of differential housing (98). Vary thickness of shims (58 – Fig. 31 and Fig. 32) to change pinion depth.

Thrust washers (56 & 70 – Fig. 31) are alike and should be 2.0 mm thick. Bushings (72) are pressed into supports (55 & 71). Inside diameter of bushing

(72) should be 75.03-75.12 mm and matching journal should be 74.97-75.00 mm. Install thrust washers (56 & 70) with chamfers toward center housing (57).

Reinstall axle as described in paragraph 5 making sure that selection of shims (54) are adjusted as outlined to provide 0.05-0.30 mm end clearance. Tighten screws (52) to 149 N·m torque.

ENGINE AND CLUTCH
FRONT SPLIT
All Models

35. It is necessary to separate the front axle, radiator and front frame from tractor for some work including service to timing gears and cover.

Detach ground cable from battery, disconnect headlight wire, disconnect hood holding bracket, then unbolt and

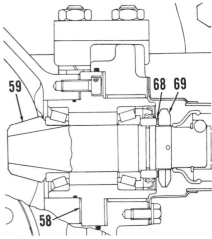

Fig. 32 – Cross-section of drive pinion (59) showing location of shims (58) used to adjust mesh position (pinion depth) and shims (68) used to permit locking nut (69) with cotter pin. Refer to Fig. 31 for exploded view.

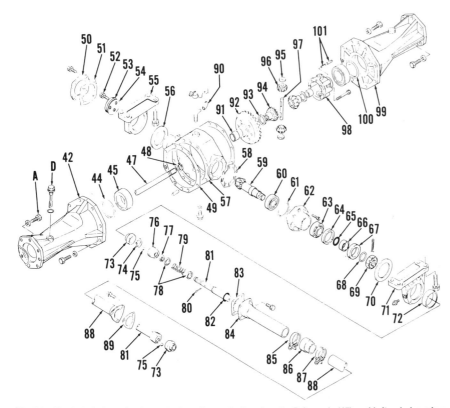

Fig. 31 – Exploded view of axle center housing and related parts. Drive axle (47) and left axle housing (42) are also shown in Fig. 30.

D. Oil level dipstick	58. Shims	71. Rear pivot support	86. Boot
43. Roll pin	59. Drive pinion	72. Bushing (2 used)	87. Clamp
44. Shim (also 100)	60. Tapered roller bearing (same as 63)	73. Rubber boot	88. Drive shaft rear cover
45. Ball bearing (carrier)		74. Snap ring	89. Gasket
47. Drive axle	61. "O" ring	75. Steel balls	90. Vent tube
48. Snap ring	62. Pinion housing	76. Coupling	91. Bushing
49. Gasket	63. Tapered roller bearing (same as 60)	77. Spline seal	92. Ring gear
50. Cover		78. Washers	93. Thrust washer
51. Gasket	64. Spacer	79. Spring	94. Side gear
52. Cap screws (3 used)	65. "O" ring	80. Cotter pin	95. Thrust washer
53. Plate	66. Wear sleeve	81. Drive shaft	96. Pinions
54. Shims	67. Oil sleeve	82. "O" ring	97. Pinion shaft
55. Front pivot support (includes 72)	68. Shims	83. Gasket	98. Differential housing
56. Thrust washer (same as 70)	69. Castellated nut	84. Drive shaft front cover	99. Right axle housing
	70. Thrust washer (same as 56)	85. Clamp	100. Shims (also 44)
57. Center housing			101. Roll Pins

Fig. 33 – View showing special gage (G) used for checking pinion depth. Refer to text.

remove hood. Disconnect battery cable from starter, remove fan belt guard and drain cooling system. Disconnect air intake hose from intake manifold (or turbocharger of models so equipped). Detach radiator overflow hose from expansion tank of 850 and 950 models. On all models, detach upper and lower radiator hoses from engine. Disconnect drain line and radiator support rod from engine block. Detach drag link from steering arm, block rear wheels and support tractor under clutch housing. Insert blocks between sides of front frame and axle to prevent tipping. On models with front wheel drive, unbolt drive shaft cover, loosen clamps then slide front cover toward rear. Disconnect spring loaded drive shaft collar, being careful not to lose balls. Rear of drive shaft can be similarly detached to remove drive shaft. On all models, support front frame and radiator assembly, unbolt frame side rails from engine, then roll front end away from tractor.

Reattach front end in reverse of separation procedure. Tighten screws which attach frame side rails to engine to 90 N·m torque.

CLUTCH SPLIT

All Models

36. To separate tractor between engine and clutch housing, refer to the following procedure.

Detach ground cable from battery, shut fuel off and disconnect fuel outlet line from filter housing. Detach filter housing from rear of engine and move filter toward rear. Detach hydraulic lines from hydraulic pump and throttle linkage from injection pump. Disconnect tachometer cable, then move cable and housing back out of the way. Disconnect interfering wires from starter, temperature sensor, thermostart, oil pressure sensor, alternator and headlights, then move wires back out of the way. Detach drag link from pitman arm and, on power steering equipped models, detach rear of cylinder from support bracket and hydraulic lines to control valve. On four wheel drive models, unbolt drive shaft cover, loosen clamps, then slide front cover toward rear. Disconnect spring loaded drive shaft collar, being careful not to lose drive balls. Rear of drive shaft can be similarly detached to remove drive shaft. On all models, block rear wheels, support tractor under clutch housing and attach movable overhead hoist to engine lifting eyes. Remove engine to clutch housing screws, then carefully roll front of tractor away from clutch housing.

When assembling, observe the following: Be sure that both return springs are attached to throw out bearing and that front and rear sections of tractor are aligned before attempting to rejoin. Engage pto and turn pto output shaft while pushing tractor together. Tighten pump lines to hydraulic pump cap screws to 8 N·m torque and clutch housing to engine cap screws to 90 N·m torque.

R&R ENGINE WITH CLUTCH

All Models

37. To remove engine and clutch assembly from tractor, first separate front axle, frame and radiator as outlined in paragraph 35, then proceed as follows:

Shut fuel off and disconnect fuel outlet line from filter housing. Detach filter housing from rear of engine and move filter toward rear. Detach hydraulic lines from hydraulic pump and throttle linkage from injection pump. Disconnect tachometer cable, then move cable and housing back out of the way. Disconnect interfering wires from starter, temperature sensor, thermostart, oil pressure sensor, alternator and headlights, then move wires back out of the way. Attach movable overhead hoist to engine lifting eyes. Remove engine to clutch housing screws, then carefully roll front of tractor away from clutch housing.

When reassembling, reverse removal procedure and observe the following. Be sure that both springs are attached to throw out bearing. Cap screws attaching hydraulic lines to pump should be tightened to 8 N·m torque, clutch housing to engine screws should be tightened to 90 N·m torque and screws attaching front frame to engine should be tightened to 90 N·m torque.

CYLINDER HEAD

All Models

40. **REMOVE AND REINSTALL.**

To remove cylinder head, refer to paragraph 45 and remove rocker arm cover. Drain coolant, remove fan belt, both radiator hoses and air inlet pipe. On models with turbocharger refer to paragraph 90 and remove turbocharger and surge tank. On all models, remove fuel leak-off and injection lines. Remove injection nozzles and intake manifold. Remove cylinder head lubrication oil line, then unbolt and remove exhaust manifold. Loosen cylinder head retaining screws in reverse of order shown in Fig. 40 or Fig. 41, then unbolt and remove cylinder head.

When reinstalling, coat each side of cylinder head gasket with "Dow Corning/RTV Silicon Gasket" or equivalent, position gasket and head on block. Coat threads with engine oil and install head retaining screws. On all models, screws should be tightened in sequence shown in Fig. 40 or Fig. 41 in three steps. On 850 and 950 models, initial torque should be 60 N·m, intermediate torque should be 120 N·m and final torque should be 160-180 N·m. On 1050 models, screws (5 & 6 – Fig. 41) are smaller and require lower torque than other screws. Three step tightening in sequence shown in Fig. 41 is required. Initial torque should be 60 N·m for all screws except (5 & 6); 27 N·m torque for screws (5 & 6). Intermediate torque should be 120 N·m for all screws except (5 & 6); 54 N·m for screws (5 & 6). Final torque should be 160-180 N·m for screws (1 through 4 and 7 through 10); 75 N·m torque for screws (5 & 6).

On all models, be sure that retaining screw final torque is even. Position wear caps on valves, then install rocker arm shaft supports and retaining nuts. Tighten shaft support nuts to 56 N·m torque on 850 models; 66 N·m torque for 950 and 1050 models. Refer to paragraph 45 for adjusting valve clearance, install rocker arm cover, then adjust decompression controls as outlined in paragraph 45. Refer to paragraph 90 for installing turbocharger on models so equipped.

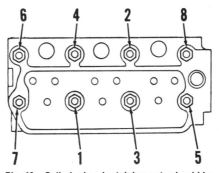

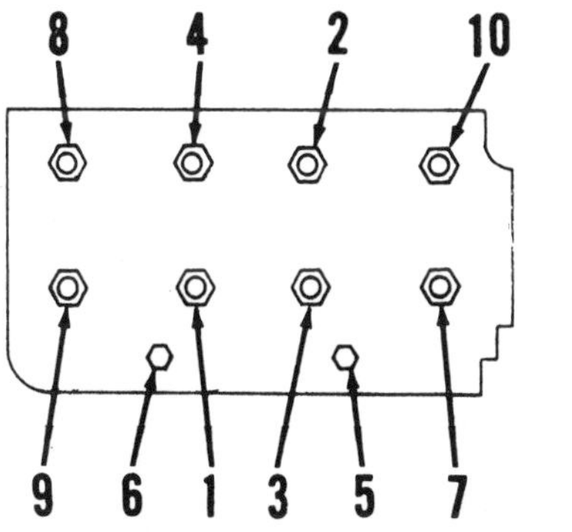

Fig. 40 – Cylinder head retaining nuts should be tightened in sequence shown for 850 and 950 models.

Fig. 41 – Cylinder head retaining nuts and cap screws should be tightened in sequence shown for 1050 models.

41. **OVERHAUL.** Refer to appropriate paragraphs 46, 47, 48 and 49 for servicing rocker arms, valves, seats, guides and springs. Check cylinder head for flatness using a straight edge and feeler gage. Head should be flat within 0.03 mm when checked across retaining bolt holes longitudinally, laterally and diagonally. Maximum warpage limit is 0.1 mm.

TAPPET GAP ADJUSTMENT

All Models

45. Recommended valve clearance (tappet gap) is 0.2 mm cold for inlet and exhaust valves of 850 models; 0.15 mm cold for inlet and exhaust valves of 950 and 1050 models. Valve clearance should be checked after assembling cylinder head to engine and approximately every 300 hours.

To remove rocker arm cover, on 1050 models, loosen clamps and remove hoses from surge tank. Unbolt and remove surge tank from rocker arm cover. On 850 and 950 models, disconnect decompression linkage and unbolt coolant expansion tank support from rocker arm cover. On all models unbolt thermostart reservoir support from rocker arm

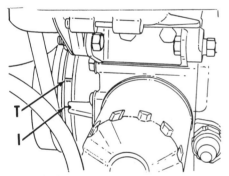

Fig. 45—Beginning of injection should occur when leading mark (I) is aligned with notch as shown. When mark "1" is aligned with notch, rear (No. 1) cylinder is at TDC.

Fig. 46—Covers (C) permit access to decompressor adjustment screws.

cover. Unbolt and remove rocker arm cover. Firing order is 1-3-2 and number 1 cylinder is at rear; number 3 cylinder is front. Check and adjust valve clearance when piston is at TDC on compression stroke. TDC mark for number "1" (rear) cylinder is marked on crankshaft pulley as shown in Fig. 45. Front (number 3) cylinder top dead center occurs 240 degrees after number "1" TDC and center (number 2) cylinder TDC is 240 degrees after front cylinder.

Reinstall rocker arm cover by reversing removal procedure. Adjust decompression controls on models so equipped as follows: Remove three covers (C – Fig.46) from top of rocker arm cover, then turn crankshaft until number "1" (rear) cylinder is at TDC on compression stroke. Push decompression lever toward rear (decompression) position. Loosen locknut (A – Fig. 47), then turn screw (B) until clearance is 0 without moving valve. Turn screw (B) in one full turn, then tighten locknut (A). This adjustment will hold valve open 0.8 mm when decompression device is engaged. Turn crankshaft 240 degrees clockwise, adjust decompression setting for number "3" (front) cylinder, then turn crankshaft 240 degrees clockwise and adjust decompression for center (No. 2) cylinder.

VALVES AND SEATS

All Models

46. Intake and exhaust valves are not interchangeable. Valve seats can be reconditioned using 15, 45 and 70

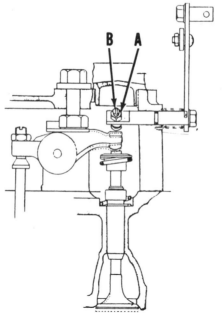

Fig. 47—Views of decompressor locknut (A) and adjusting screw (B). Cross-section drawing shows method of operation.

degree stones. Valve face and seat angles for both inlet and exhaust is 45 degrees. Recommended seat width is 2.12 mm. Seats can be narrowed using 15 and 70 degree stones. Valve is recessed 1.15 mm below gasket surface of head when new and should not be recessed more than 1.65 mm. Renew valve if margin is less than 0.5 mm. Valve head diameter for 850 models is 33.5 mm for intake, 27.5 mm for exhaust; for 950 and 1050 models intake valve head diameter is 38.5 mm and exhaust valve head diameter is 32.5 mm. Valve stem diameter is 6.95-6.96 mm for 850 models when new; minimum limit is 6.90 mm. Valve stem diameter is 7.96-7.97 mm for 950 and 1050 models when new; minimum limit is 7.90 mm. Install new valve stem seals when ever valves are serviced. Valve seat inserts are used on all valve seats of 1050 models.

VALVE GUIDES

All Models

47. Inlet and exhaust valve guides are not interchangeable. Passage end of exhaust guide is counterbored, intake guide is not. Stem to guide clearance should be 0.04-0.06 mm for both inlet and exhaust valves of all models and wear limit is 0.15 mm. Manufacturer recommends knurling worn guides if clearance is less than 0.2 mm.

Valve guide ID new is 7.00-7.02 mm for 850 models, 8.01-8.03 mm for 950 and 1050 models. New guides should be pressed into cylinder head until groove

in guide is just above upper surface of cylinder head.

VALVE SPRINGS

All Models

48. Valve springs are interchangeable for inlet and exhaust valves. Renew springs which are distorted, heat discolored or fail to meet test specifications which follow:

850 Models
Free Length, New . . . . 41.0 mm
   Minimum Limit . . . . 40.7 mm
Test Pressure N @ mm,
   New . . . . 160.83 @ 30.73
      Minimum Limit . . . . 154.95 N
Maximum Distortion (tilt) . . . . . 1.43 mm
950 & 1050 Models
Free Length, New . . . . 40.0 mm
   Minimum Limit . . . . 39.7 mm
Test Pressure N @ mm,
   New . . . . 168.67 @ 32.26
      Minimum Limit . . . . 161.81 N
Maximum Distortion (tilt) . . . . . 1.41 mm

ROCKER ARMS AND PUSH RODS

All Models

49. Inlet and exhaust rocker arms are interchangeable, but front cylinder rocker arms are offset differently than rear two cylinders. To remove rocker arms, refer to paragraph 45 and remove rocker arm cover. Rocker arm bushings are not available separately from rocker arms. Bushing ID is 17.02-17.03 mm new, with wear limit of 17.10 mm. Shaft is integral with rocker shaft support and should be 16.98-17.00 mm diameter, with minimum diameter of 16.90 mm. Bushing to shaft clearance should be 0.016-0.052 mm, with wear limit of 0.15 mm. Refer to paragraph 45 for adjustment of valve clearance (tappet gap) and adjustment of decompression controls.

Push rods should be straight and can be checked by rolling on flat plate.

Fig. 52 — View of timing marks on camshaft and crankshaft gears.

Length of new push rod is 196 mm. Renew push rod if less than 194.0 mm long or if bent more than 0.3 mm.

CAM FOLLOWERS

All Models

50. All models are equipped with mushroom type cam followers which can only be removed from below after removing camshaft as outlined in paragraph 55 and oil pan. Check surface of cam follower which contacts camshaft and renew followers if worn. Also check camshaft carefully if follower shows wear and renew camshaft if chipped, broken, scored or otherwise worn. Cam follower OD of 11 mm should have 0.01-0.03 clearance in block bore. Cam follower wear limit is 10.95 mm and follower to bore clearance should not exceed 0.10 mm.

VALVE TIMING

All Models

52. Refer to paragraph 53 for removal of timing gear cover. Valves are properly timed when timing marks are aligned as shown in Fig. 52. Marks align every second revolution of crankshaft. Hydraulic pump drive gear and engine lubrication pump drive gear do not need to be timed.

TIMING GEAR COVER

All Models

53. To remove the timing gear cover, first perform front split as outlined in paragraph 35. Remove fan belt, fan, hydraulic pump and crankshaft pulley. Disconnect tachometer drive cable from housing. Remove injection pump chamber cover plate, then disconnect governor spring (1 – Fig. 53), remove pin (2) and detach governor link from rack. Be careful not to deform spring when detaching spring from speed control lever. Unbolt and remove timing gear cover.

When reinstalling cover, be sure not to bend or otherwise damage governor link (16 – Fig. 54). Refer to paragraphs 83 and 84 for governor adjustments if correct adjustment is questioned or if any adjustment point (3, 4, 5, 6 or 7) has been disturbed. Be sure that sleeve (17) correctly engages weights (20) when installing cover. Both sides of timing gear cover gasket should be coated with "Permatex No. 1" or equivalent. Tighten nut which has left hand thread and retains governor weight support (21) to camshaft to 80 N·m. Screws attaching timing gear cover to block should be tightened to 25-27 N·m torque and crankshaft pulley retaining screw should be tightened to 65 N·m torque.

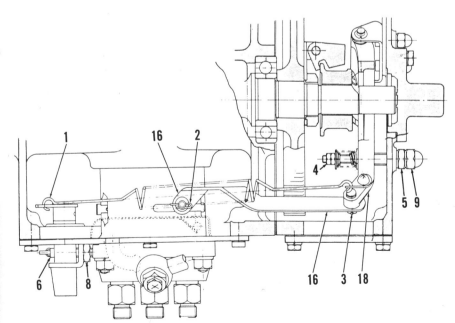

Fig. 53 — Cross-section of governor assembly. Refer to text.

1. Governor spring	4. Adjusting nuts	9. Lock cap nut
2. Clip	5. Locknut	16. Governor link
3. Adjusting nut	6. High speed stop screw	18. Governor lever
	8. Locknut	

TIMING GEARS

All Models

54. Refer to paragraph 53 for removal of timing gear cover. Backlash between crankshaft and camshaft should be 0.08-0.16 mm with wear limit of 0.30 mm. Hydraulic pump drive gear and engine oil pump drive gear do not need to be timed. Backlash between engine oil pump gear and crankshaft gear should be 0.1-0.3 mm with wear limit of 0.6 mm.

Crankshaft gear can be pulled from crankshaft using threaded holes in gear and a suitable puller. Align keyway with key and timing marks (Fig. 52), then use threaded hole in crankshaft to push gear onto crankshaft.

Refer to paragraph 55 to remove camshaft if camshaft gear is to be removed.

CAMSHAFT

All Models

55. To remove the camshaft, first remove cylinder head as outlined in paragraph 40, fuel injection pump as outlined in paragraph 68, then refer to paragraph 53 and remove timing gear cover. Use magnetic holding tools to raise cam followers away from camshaft. Remove camshaft bearing retainer screw (Fig. 55), then pull camshaft and gear forward out of cylinder block. Camshaft gear and front bearing can be pressed from camshaft after removing nut (left hand threads) and governor weight assembly. Measure camshaft lobes and journals and compare with the following specifications:

Rear (Flywheel) Journal-
Diameter, new . . . . . . 33.92-33.95 mm
Wear limit . . . . 33.81 mm
Rear Bushing-I.D.
New . . . . 34.00-34.02 mm
Wear limit . . . . 34.07 mm

Center Journal-
Diameter, new . . . . . . 45.92-45.95 mm
Wear limit . . . . 45.87 mm
Center Bushing-I.D.
New . . . . 46.00-46.02 mm
Wear limit . . . . 46.07 mm
Inlet and Exhaust Lobe-
Height, new . . . . 38.60 mm
Wear limit . . . . 38.10 mm
Fuel Pump Lobe-
Height, new . . . . 35.00 mm
Wear limit . . . . 34.90 mm

Press front bearing on camshaft, position key in slot, then press camshaft gear onto camshaft. Install governor weights and support, then install nut with left hand thread to final torque of 80 N·m. Position camshaft in cylinder block bores, then install front bearing retainer and screw. Tighten retaining screw to 20 N·m torque.

To remove cam followers, remove oil pan, then the mushroom type cam followers can be withdrawn from below. Refer to paragraph 50 for specifications.

Refer to paragraph 68 for installing fuel injection pump, paragraph 53 for installing timing gear cover and paragraph 40 for installing cylinder head.

ROD AND PISTON UNITS

All Models

56. Connecting rod and piston units are removed from above after removing cylinder head as outlined in paragraph 40 and removing oil pan. Remove connecting rod cap and push piston and rod assembly up out of cylinder block.

When assembling, space ring end gaps at 90 degree intervals from each other and 45 degrees from piston pin centerline. Coat pistons, liners and ring compressor with clean engine oil. Identification marks on connecting rods should be toward rear (flywheel) end of engine. Tighten connecting rod cap retaining screws to 55 N·m for 850

models, 65 N·m for 950 and 1050 models.

PISTONS, SLEEVES AND RINGS

All Models

57. Measure stand out of cylinder liners (sleeves) above cylinder block. A puller should be used to pull wet type cylinder liners from block bores. Be sure that manufacturers marks on all rings are toward top when assembling rings to pistons. Check clearances against the values which follow:

850 Models
Liner Stand Out . . . . 0.05-0.13 mm
Cylinder Liner I.D. –
New . . . . 80.00-80.03 mm
Wear Limit . . . . 80.20 mm
Piston Skirt to Cylinder Liner-
Clearance New . . . . 0.08-0.15 mm
Wear Limit . . . . 0.30 mm
Piston Skirt Diameter at Bottom
of Skirt/Right Angles to Piston Pin-
New . . . . 79.86-79.90 mm
Wear Limit . . . . 79.78 mm
Ring End Gap
All Rings, desired . . . . . . 0.3-0.5 mm
Wear Limit . . . . 1.5 mm
Ring to Groove Side
Clearance-Wear Limit
Top Ring . . . . 0.30 mm
Second, third & oil rings . . . . 0.25 mm

950 and 1050 Models
Liner Stand Out . . . . 0.07-0.15 mm
Cylinder Liner I.D.-
New . . . . 90.00-90.03 mm
Wear Limit . . . . 90.20 mm
Piston Skirt to Cylinder Liner-
Clearance New . . . . 0.17-0.24 mm
Wear Limit . . . . 0.50 mm
Piston Skirt Diameter at Bottom
of Skirt/Right Angles to Piston Pin-
New . . . . 89.86-89.89 mm
Wear Limit . . . . 89.78 mm
Ring End Gap
All Rings, desired . . . . 0.3-0.5 mm
Wear Limit . . . . 1.5 mm

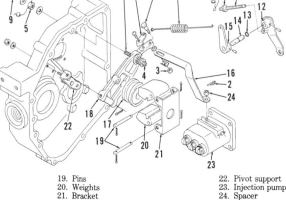

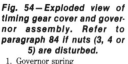

Fig. 54 — Exploded view of timing gear cover and governor assembly. Refer to paragraph 84 if nuts (3, 4 or 5) are disturbed.

1. Governor spring
2. Clip
3. Adjusting nut
4. Adjusting nuts
5. Locknut
6. High speed stop screw
7. Torque spring and shaft assembly
8. Locknut
9. Lock cap nut
10. Attaching screw
11. Spring tab
12. External speed control lever
13. "O" ring
14. Shaft
15. Internal speed control lever
16. Governor link
17. Governor sleeve
18. Governor lever
19. Pins
20. Weights
21. Bracket
22. Pivot support
23. Injection pump
24. Spacer

Fig. 55 — View of governor weights and bracket assembled on cam gear. Screw identified by arrow retains camshaft front bearing.

Ring to Groove Side Clearance-
Wear Limit
    Top Ring (950 Models) . . . . . .0.30 mm
    Top Ring (1050 Models) . . . . *
    Second, third & oil rings
      (950 & 1050 Models) . . . . . .0.25 mm
*The top ring on 1050 models is semi-keystone type. Measure groove clearance with groove clean and outside diameter of ring flush with ring lands. Maximum groove clearance limit is 0.20 mm.

Be sure that liner bore in cylinder block is clean, then install "O" rings in bore grooves. Lubricate "O" rings and cylinder liner with soap lubricant, then press cylinder firmly into bore until seated. If stand out exceeds 0.13 mm for 850 models, 0.15 mm for 950 and 1050 models, remove cylinder liner and check for rust, scale or burrs that prevent seating.

PISTON PIN

All Models

58. Piston pin is interference fit in piston bores and should have 0.02-0.05 mm clearance in rod bushing. Remove pin retaining snap rings from piston, heat piston to 80 degrees C., then press piston pins from piston bores. Refer to the following specification data:

850 Models
Pin Diameter-
    New . . . . .25.99-26.00 mm
    Wear Limit . . . . .25.90 mm
Bushing Diameter-
    New . . . . .26.02-26.04 mm
    Wear Limit . . . . .26.10 mm

950 and 1050 Models
Pin Diameter-
    New . . . . .29.991-30.000 mm
    Wear Limit . . . . .29.90 mm
Bushing Diameter
    New . . . . .30.025-30.038 mm
    Wear Limit . . . . .30.10 mm

Align both holes in bushing with holes in connecting rod. When assembling pistons, pins and connecting rods, observe the following:
Install one piston pin retaining snap ring in one groove of each piston, then heat pistons to 80 degrees C. Coat piston pins and rod bushings with clean engine oil and install cool pin through rod bushing and heated piston. Be sure to install second retaining ring in piston groove. After cooling, connecting rod should move smoothly on piston pin.

CONNECTING RODS AND BEARINGS

All Models

59. Connecting rod crankpin bearing inserts are available in standard size only. Rod bearing inserts can be installed from below after removing oil pan. Refer to the following specification data:

850 Models
Piston Pin Bushing I.D.-
    New . . . . .26.02-26.04 mm
    Wear Limit . . . . .26.10 mm
Crankpin Journal Diameter-
    New . . . . .49.95-49.96 mm
    Wear Limit . . . . .49.92 mm
Connecting Rod Bearing to
Crankpin Clearance-
    New . . . . .0.04-0.09 mm
    Wear Limit . . . . .0.15 mm
Rod Bolt Torque . . . . .55 N·m

950 and 1050 Models
Piston Pin Bushing I.D.-
    New . . . . .30.025-30.038 mm
    Wear Limit . . . . .30.10 mm
Crankpin Journal Diameter-
    New . . . . .53.95-53.96 mm
    Wear Limit . . . . .53.92 mm
Connecting Rod Bearing to
Crankpin Clearance-
    New . . . . .0.04-0.09 mm
    Wear Limit . . . . .0.15 mm
Rod Bolt Torque . . . . .65 N·m

CRANKSHAFT AND MAIN BEARINGS

All Models

60. The crankshaft is supported in four main bearings. To remove crankshaft and/or main bearings, proceed as follows: Separate the front axle, frame and radiator as outlined in paragraph 35, then refer to paragraph 37 to complete removal of engine. Refer to paragraph 40 to remove cylinder head, paragraph 53 to remove timing gear cover, paragraph 56 to remove rod and piston units and paragraph 62 for removing clutch and flywheel. Pull crankshaft gear from crankshaft and remove oil pump pick up tube. Turn engine so that rear of crankshaft is up, then remove rear main bearing and seal cap. Jack screw (threaded) holes are provided in rear cap to facilitate removal. Attach hoist to crankshaft and raise crankshaft slightly to remove weight from center main bearings. Remove center main bearing set bolts, then carefully lift crankshaft and center main bearing housings out of cylinder block. Remove bolts from center main bearing housings, then separate and remove housing and bearing inserts from

crankshaft. Inspect main bearings and crankshaft using the following specification data:

850 Models
Crankshaft Main Journals-
    Diameter, New . . . . . .69.95-69.96 mm
    Wear Limit . . . . .69.92 mm
Main Bearing Diameters-
    New . . . . .70.00-70.04 mm
    Wear Limit . . . . .70.12 mm
Main Journal to Bearing Clearance-
    New . . . . .0.04-0.10 mm
    Wear Limit . . . . .0.15 mm

950 and 1050 Models
Crankshaft Main Journals-
    Diameter, New-Front &
      Center . . . . .69.95-69.96 mm
    Diameter, new-Rear .89.95-89.96 mm
    Wear limit-Front & Center .69.92 mm
    Wear Limit-Rear . . . . .89.92 mm
Main Bearing Diameters-
    New-Front & Center .70.00-70.04 mm
    New-Rear . . . . .90.00-90.04 mm
    Wear limit-Front & Center .70.12 mm
    Wear limit-Rear . . . . .90.15 mm
Main Journal to Bearing Clearance-
    New-Front & Center . . .0.04-0.10 mm
    New-Rear . . . . .0.07-0.13 mm
    Wear Limit-Front &
      Center . . . . .0.15 mm
    Wear Limit-Rear . . . . .0.20 mm

Front main bearing (6 – Fig. 60) is pressed into cylinder block and rear main bearing (11) is pressed into rear housing (12). Be sure to align oil holes in bearings with passages in block and housing when pressing new bearings into position. Rear seal (13) should be pressed into rear bore of housing (12) with lip toward inside (front).

Be sure that bearing tangs on intermediate bearings (7 & 9) correctly engage notches in housing (8). Thrust bearing (9) should be installed at third (from front) main journal and casting mark "F" on housing (8) should be toward **FLYWHEEL**. Tighten screws holding halves of housings together to 65-70 N·m torque.

Be sure that all main bearings and seals are lubricated with engine oil and stand cylinder block on front face so that crankshaft and intermediate housings can be inserted from above. Lower crankshaft, intermediate main bearings and housings into block aligning holes in intermediate housings with bolt holes in block (crankcase). Install both intermediate housing retaining screws loosely, then tighten screw retaining thrust main bearing housing to 80 N·m torque. Tighten screw retaining second (from front) main bearing housing to 80 N·m torque. Make sure that crankshaft rotates smoothly, then install rear main bearing and seal housing. On 850 models make sure that manufacturer's casting mark is toward bottom. On all 950 and 1050 models, bolt holes are not symmetrical and installation is possible in only one position. Locating casting mark on 850 models to bottom will align oil holes in block and bearing housing. On all models, rear main bearing housing retaining screws should be tightened to 43-49 N·m torque. Flywheel retaining screws should be renewed if removed for any reason.

Reverse disassembly procedure to complete assembly.

CRANKSHAFT REAR OIL SEAL

All Models

61. The lip type crankshaft rear oil seal (13 – Fig. 60) is contained in rear housing (12) which also contains the rear main bearing. Seal can be renewed after splitting tractor between engine and clutch housing as outlined in paragraph 36, then removing clutch, flywheel and housing.

Lip of rear seal should be toward in-side (front). Manufacturer's casting mark on rear housing of 850 models should be centered toward bottom so that oil holes in block and in housing will be aligned. Attaching screws are not symmetrical permitting installation of rear housing in only one position on 950 and 1050 models. On all models, rear housing attaching screws should be tightened to 43-49 N·m torque. Flywheel retaining screws should be renewed if removed.

FLYWHEEL

All Models

62. Flywheel is positively located on crankshaft by a dowel pin and six screws. Make sure that mating surfaces of flywheel and shaft are clean and free of dirt, rust and burrs. Flywheel retaining screws should be renewed if removed. Correct torque for new screws is 75 N·m.

Pilot bearing is bushing type pressed into flywheel of 850 and 950 models. Inside diameter of bushing should be 14.96-14.98 mm and OD of clutch shaft should be 14.935-14.953 mm. Pilot journal to bushing clearance should be less than 0.3 mm. Old bushing can be pulled from bore with blind hole puller. Install new bushing flush with recess in flywheel.

Pilot bearing is ball type on 1050 models and should be renewed if roughness is noticed.

On 850 and 950 models clutch to flywheel retaining screws should be tightened to 27 N·m torque. Clutch to flywheel retaining screws or 1050 models should be tightened to 22 N·m torque.

OIL PUMP AND RELIEF VALVE

All Models

63. The rotor type oil pump is attached to front face of engine block and is driven by gear on front of engine crankshaft. Backlash between oil pump gear and crankshaft gear should be 0.1-0.3 mm with wear limit of 0.6 mm.

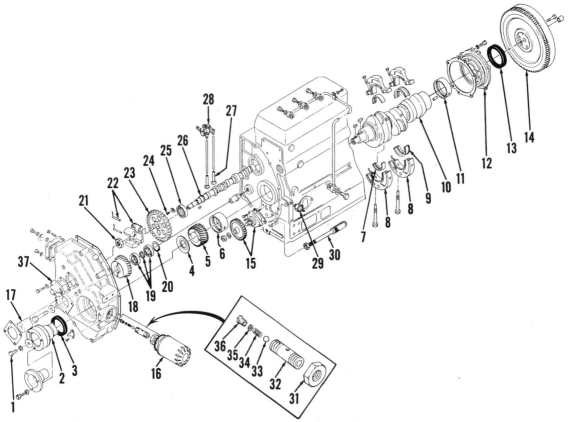

Fig. 60 – Partially exploded view of timing gears, camshaft and crankshaft. Inset is oil pressure relief valve.

1. Screw	12. Rear main bearing housing	25. Camshaft front bearing
2. Crankshaft pulley	13. Rear oil seal	26. Camshaft
3. Front oil seal	14. Flywheel	27. Cam follower
4. Oil deflector	15. Oil pump	28. Rocker arms
5. Crankshaft gear	16. Oil filter	29. Oil pressure sender
6. Front main bearing	17. Hydraulic pump drive coupling	30. Oil pick up tube
7. Front intermediate main bearing	18. Pump drive gear	31. Nut
8. Center main bearing housings	19. Bearings and spacer	32. Housing
9. Thrust bearing	20. Snap ring	33. Relief valve ball
10. Crankshaft	21. Nut	34. Spring
11. Rear main bearing	22. Governor weight	35. Shims
	23. Camshaft gear	36. Retainer
	24. Retaining screw	37. Tachometer drive

Pump can be removed after removing timing gear cover as outlined in paragraph 53. Pump rotor end play in housing can be measured with straightedge across pump body and inserting feeler gage between straightedge and pump rotor. End play should be 0.01-0.05 mm and should not exceed 0.10 mm. Pump outer rotor to body bore should be 0.05-0.10 mm with wear limit of 0.15 mm. Pump inner rotor tip to outer rotor lobe clearance should be 0.05-0.10 mm with wear limit of 0.15 mm. Oil pump gear retaining nut should be tightened to 16-20 N·m torque. Be careful not to overtighten. Oil pressure sender is located at (29 – Fig. 60). Coat threads of oil pick-up tube (30) with "Loctite" or equivalent, screw pipe into block 6-7½ turns until opening is down, then tighten locknut.

Oil pressure relief valve (33, 34, 35 & 36) is located in body (32). Correct oil pressure is 2.45-3.92 bar (36-57 psi). Addition or removal of one shim (35) will change oil pressure approximately 0.17 bar (2.6 psi).

DIESEL FUEL SYSTEM

All models are equipped with a Yanmar PFK3K three plunger injection pump, three Yanmar YDN-OSYD1 injection nozzles and swirl type precombustion chambers.

Because of extremely close tolerances and precise requirements of all diesel components, it is of utmost importance that clean fuel and careful maintenance be practiced at all times. Unless necessary special tools are available, service on injectors and injection pump should be limited to removal, installation and exchange of complete assemblies. It is impossible to re-calibrate an injection pump or reset an injector without proper specifications, equipment and training.

FUEL FILTERS AND BLEEDING

All Models

65. **OPERATION AND MAINTENANCE.** The fuel system includes a fuel filter that should be checked every 10 hours of operation, cleaned every 100 hours and renewed whenever plugged with contaminants (or at least once each year). Fuel filter housing incorporates a shut-off valve.

The fuel system should be bled if fuel tank has been allowed to run dry, if fuel filter, lines or other components within the system has been disconnected or removed or if engine has not been operated for long period of time. If the engine fails to start or if it starts then stops, the cause could be air in the system which should be removed by bleeding. Two bleed screws are located at top of fuel filter housing and an additional bleed screw is located on fitting of fuel pump inlet line.

Partially open throttle and attempt to start engine. If engine does not start, loosen high pressure fuel line at each injector and continue to crank engine until fuel escapes from loosened connection. Tighten compression nuts and start engine.

INJECTION PUMP

All Models

67. **TIMING TO ENGINE.** Beginning of injection should occur at 24-28 degrees BTDC for 850 models, 23-27 degrees BTDC for 950 and 1050 models. Proceed as follows to time injection pump to engine.

Shut fuel off at filter and disconnect fuel delivery tube (27 – Fig. 66) for rear (number 1) cylinder. Unscrew delivery valve holder (26) from pump, remove delivery valve (23) and spring (24), then reinstall delivery valve holder (26). Attach timing fixture (tube) to number one outlet of pump with outlet of fixture toward rear. Pull out decompression knob, then use a 19 mm socket to rotate engine crankshaft slowly in normal clockwise direction until the rear (number 1) piston begins coming up on compression stroke. Turn fuel on and continue turning crankshaft until fuel just stops flowing from timing fixture. Point at which fuel just stops flowing is beginning of injection and should occur as mark on crankshaft pulley just aligns with notch as shown in Fig. 67. Flow of fuel from delivery valve holder (and timing fixture) will change from steady stream to drops, then stop (which indicates beginning of injection). If fuel does not stop flowing when marks (Fig. 67) are aligned, piston may be on exhaust stroke. If mark is past pointer at beginning of injection, remove some shims (2 – Fig. 66) from between injection pump and engine block. If mark has not yet reached point at beginning of injection, add shims (2) between injection pump and engine block. Changing shim thickness 0.1 mm will change timing approximately 1 degree. After timing is correct, remove timing fixture and delivery valve holder. Install delivery valve, spring and delivery valve holder using new "O" ring (25) and copper ring (22). Tighten delivery valve holder to 40-45 N·m torque.

68. **REMOVE AND REINSTALL.** To remove injection pump, clean area around pump and lines, turn fuel off at filter, then disconnect inlet line to pump and outlet (discharge) lines from pump. Be sure to cover all openings in pump and lines immediately to prevent entrance of dirt. Detach speed control rod from lever. Remove covers from pump chamber (37 – Fig. 66) and governor chamber (on timing gear cover). Disconnect governor spring (1 – Fig. 68) from speed control internal lever and clip (2) from pin on pump control rack. Detach governor link (16) from vertical pin, then unbolt and remove pump from engine. Be careful not to lose shims (2 – Fig. 66) which are used to set injection pump timing.

Further disassembly of pump is not recommended.

When installing, reverse removal procedure, using same thickness of shims (2) as were removed. Check pump timing as outlined in paragraph 67 and vary thickness of shims (2) as required to provide correct timing.

INJECTOR NOZZLES

All Models

70. **TESTING AND LOCATING A FAULTY NOZZLE.** If engine is missing and fuel system is suspected as being the cause of trouble, system can be checked by loosening each injector line connection in turn, while engine is running at slow idle speed. If engine operation is not materially affected when injector line is loosened, that cylinder is missing. Remove and test (or install a new or reconditioned unit) as outlined in appropriate following paragraphs.

71. **REMOVE AND REINSTALL.** Before removing an injector or loosening injector lines, thoroughly clean injector, lines and surrounding area using compressed air and a suitable solvent. Remove high pressure line leading from pump to injector unit and disconnect the bleed line (39 – Fig. 66) from injector by removing banjo bolt (38). Unbolt and remove injector hold down clamp then pull injector from cylinder head using a puller if necessary.

Clean exterior surface of nozzle assembly and bore in cylinder head, being careful to keep all surfaces free of dust, lint and other small particles until nozzle is installed. Tighten nozzle hold down nuts evenly to 20 N·m torque. Be sure that both nuts are tightened evenly. High pressure injector line fittings should be tightened to 27 N·m torque. Refer to paragraph 65 for bleeding system.

72. **TESTING.** A complete job of testing and adjusting the injector requires use of special test equipment. Only clean, approved testing oil should be

used in tester tank. Nozzle should be tested for opening pressure, seat leakage, back leakage and spray pattern. When tested, nozzle should open with a high-pitched buzzing sound, and cut off quickly at end of injection with a minimum of seat leakage and a controlled amount of back leakage.

Before conducting test, operate tester lever until fuel flows, then attach injector. Close valve to tester gage and pump tester lever a few quick strokes to be sure nozzle valve is not stuck, and that possibilities are good that injector can be returned to service without disassembly.

WARNING: Fuel leaves injector nozzle with sufficient force to penetrate the skin. Keep exposed portions of your body clear of nozzle spray when testing.

73. OPENING PRESSURE. Open valve to tester gage and operate tester lever slowly while observing gage reading. Opening pressure should be 157 bar (2275 psi).

Opening pressure is adjusted by adding or removing shims in shim pack (3–Fig. 73). Each shim will change opening pressure approximately 7-10 bar (100-142 psi).

74. SPRAY PATTERN. Spray pattern should be well atomized and slightly conical, emerging in a straight axis from nozzle tip. If pattern is wet, ragged or intermittent, nozzle must be overhauled or renewed.

75. SEAT LEAKAGE. Wipe nozzle

tip dry with clean blotting paper; then, operate tester handle to bring gage

Fig. 67 — View of mark indicating injection timing aligned with notch. TDC mark for number 1 cylinder is also shown.

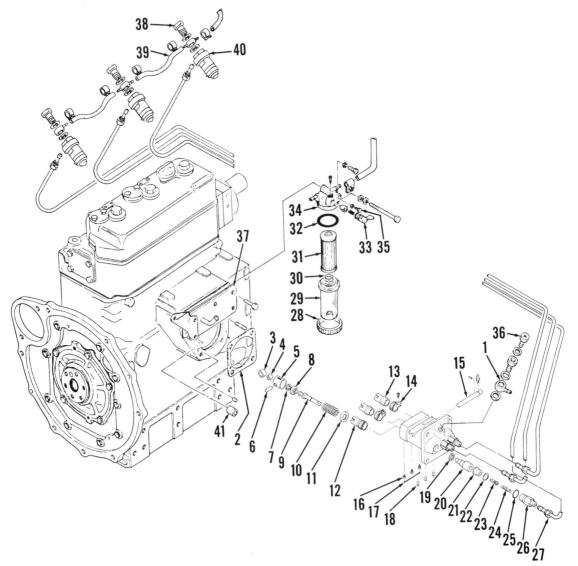

Fig. 66 — Exploded view of injection pump and filter. Pump should not be disassembled by untrained personnel or without proper equipment.

1. Inlet banjo fitting	14. Fuel control pinion (front & center)	28. Retainer
2. Timing shims	15. Rack	29. Bowl
3. Roller	16. Guide stopper	30. Spring
4. Bushing	17. Stopper pin	31. Filter element
5. Guide	18. Plunger barrel pin	32. Gasket
6. Pin	19. Gasket	33. Shut-off valve
7. Shim	20. Barrel	34. Bracket
8. Retainer	21. Delivery valve seat	35. Bleed screw
9. Plunger	22. Gasket	36. Bleed screw
10. Spring	23. Delivery valve	37. Cover
11. Spring retainer	24. Valve spring	38. Banjo fitting
12. Fuel control pinion (rear) and sleeve	25. "O" ring	39. Bleed off line
13. Sleeve (front & center)	26. Delivery valve holder	40. Injector assembly
	27. Delivery tube	41. Speed control shaft bushing

pressure to 137 bar (1990 psi) and hold this pressure for 10 seconds. If any fuel appears on nozzle tip, overhaul injector as outlined in paragraph 75.

76. OVERHAUL. Hard or sharp tools, emery cloth, grinding compound or other than approved solvents or lapping compounds must never be used.

Wipe all dirt and loose carbon from exterior of nozzle and holder assembly. Refer to Fig. 73 for exploded view and proceed as follows:

Secure nozzle in a soft jawed vise or holding fixture and remove nuts (1 & 9).

Place all parts in clean calibrating oil or diesel fuel as they are removed, using a compartmented pan and using extra care to keep parts from each injector together and separate from other units which are disassembled at the time.

Clean exterior surfaces with a brass wire brush, soaking in an approved carbon solvent if necessary, to loosen hard carbon deposits. Rinse parts in clean diesel fuel or calibrating oil immediately after cleaning to neutralize the solvent and prevent etching of polished surfaces.

Clean nozzle spray hole from inside us-

ing a pointed hardwood stick or wood splinter. Scrap carbon from pressure chamber using hooked scraper. Clean valve seat using brass scraper, then polish seat using wood polishing stick and mutton tallow.

Reclean all parts by rinsing thoroughly in clean diesel fuel or calibrating oil and assemble while parts are immersed in cleaning fluid. Make sure adjusting shim pack is intact. Tighten nozzle retaining nut (9) to a torque of 90-100 N·m and nut (1) to 70-80 N·m torque. Do not overtighten, distortion may cause valve to stick and no amount of overtightening can stop a leak caused by scratches or dirt. Retest assembled injector as previously outlined.

THERMOSTART

All Models

80. The thermostart plug is installed in intake manifold and warms inlet air by igniting small quantities of diesel fuel in the inlet manifold. Operation can be checked by viewing in intake manifold with key turned to thermostart position. After about five seconds inner coil should glow bright red. After about ten seconds, burning fuel should begin to drip from plug. If heating coils do not get hot, check for battery current at thermostart terminal. If tests indicate current is available at terminal, but coil does not get hot, disconnect wire from

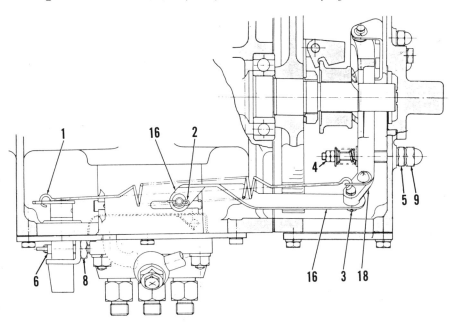

Fig. 68 — Cross-section of governor assembly. Refer to text.

1. Governor spring
2. Clip
3. Adjusting nut
4. Adjusting nuts
5. Locknut
6. High speed stop screw
8. Locknut
9. Lock cap nut
16. Governor link
18. Governor lever

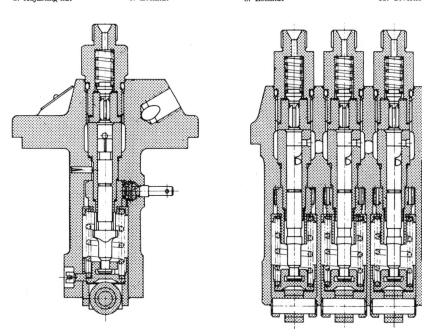

Fig. 69 — Cross-section of injection pump assembly. Disassembly should not be attempted without proper equipment and specialized training.

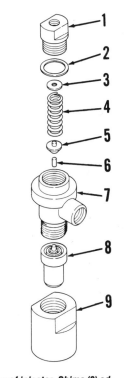

1. Cap
2. Gasket
3. Shims
4. Spring
5. Spring seat
6. Inner spindle
7. Body
8. Nozzle
9. Nut

Fig. 73 — Exploded view of injector. Shims (3) adjust opening pressure.

terminal and check continuity between terminal on thermostart plug and engine ground. Open circuit indicates damaged heating element. Air pressure of 1.4 bar (20 psi) should not cause thermostart valve to leak when checked as shown in Fig. 80.

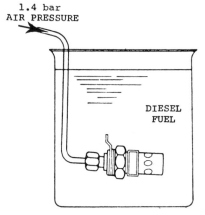

Fig. 80 — Thermostart valve can be checked for leakage as shown. Clean or renew valve if bubbles are visible from valve.

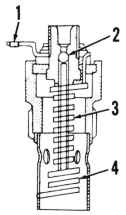

Fig. 81 — Cross-section of thermostart assembly.
1. Electrical terminal
2. Valve ball
3. Heater coil
4. Igniter coil

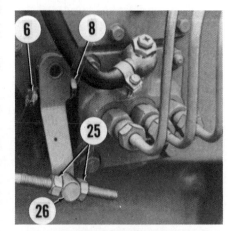

Fig. 83 — High speed is adjusted at stop screw (6) and locked by nut (8). Slow idle speed is changed by relocating pin (26) on shaft and positioning with nuts (25).

GOVERNOR

82. The governor is a variable speed flyweight type, with weights located at front of camshaft gear and control arm located inside timing gear cover. Refer to paragraph 53 for timing gear cover removal procedure and to Fig. 53 and Fig. 54 for views of governor parts.

ADJUSTMENT

All Models

83. High idle no load speed should be limited to 2725-2775 rpm for 850 models, 2575-2625 rpm for 950 and 1050 models. High idle no load speed is adjusted by stop screw (6 – Fig. 83) and adjustment is locked by nut (8). Adjustment is sealed with wire and lead seal. High idle speed of 2750 rpm is 654-666 rpm at pto for 850 models, high idle speed for 950 and 1050 models (2600 rpm) is 618-630 rpm when measured at pto shaft.

Low idle engine speed should be 850 rpm and is adjusted by turning nuts (25) to move pin (26) on throttle rod. Be sure that nuts (25) are tightened when adjustment is correct.

If normal high idle and low idle speeds can not be set satisfactorily in conventional way, refer to paragraph 84 for initial settings and adjustments.

84. Several adjustments must be accomplished while assembling the governor controls to permit correct governor action.

Torque springs, shaft (7 – Fig. 84), spring retainers and nuts (4) are available only as a complete pre-adjusted assembly. Initial setting (A) is approximately 12 mm but adjustment of nuts (4) should not be attempted.

If torque spring and shaft assembly (7) is installed in timing gear cover, assemble and connect all governor and injection pump parts. Disconnect rear end of spring (1) and loosen adjustment nut (3). Turn torque spring shaft (7) into (toward camshaft gear) timing gear cover far enough that governor lever (18) can not contact torque spring retainer even when forced. Move link (16) toward rear as far as possible, then tighten and lock adjustment nut (3). Move link (16) forward until punch mark (B) on pump control rack is centered under machined face of housing as shown, then turn torque shaft (7) out until the torque spring retainer just touches governor lever (18). Lock position with nut (5) and cap nut (9).

Complete adjustment by attaching spring (1). Check injection pump timing as outlined in paragraph 67 and engine speed as in paragraph 83.

TURBOCHARGER

The engine used in 1050 models is equipped with an exhaust driven, IHIRHB6 turbocharger. Lubrication and cooling is provided by engine oil. After the engine is operated under load, the turbocharger should be allowed to cool by idling for a

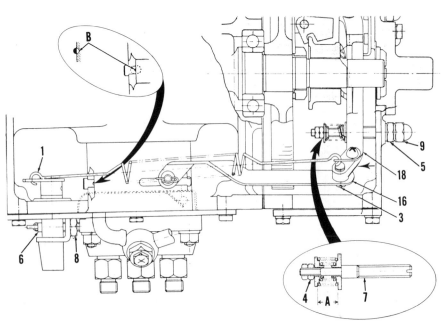

Fig. 84 — Cross-section of governor assembly. Refer to text for adjustment procedures.

A. Setting distance for torque springs
B. Mark on rack aligned with machined surface of pump body

1. Governor spring
3. Adjusting nut
4. Adjusting nuts

5. Locknut
6. High speed stop screw
7. Torque spring shaft
8. Locknut

9. Lock cap nut
16. Governor link
18. Governor lever

few minutes. **The engine should be immediately restarted if it is killed while operating at full load. When servicing the turbocharger, extreme care must be taken to avoid damaging any moving parts.**

TURBOCHARGER UNIT

1050 Models

90. **REMOVE AND REINSTALL.** Clean exterior of turbocharger and surrounding area. Unbolt exhaust outlet elbow from turbocharger, detach air inlet hose from turbocharger inlet and remove hose from turbocharger to surge tank. Detach oil inlet and return lines and cover openings to prevent entrance of dirt. Unbolt turbocharger from exhaust manifold, then lift turbocharger carefully from engine.

Oil passage in turbocharger should be filled with clean engine oil before installing. Rotate turbine shaft by hand to be sure that all surfaces of bearings and seals are adequately lubricated. Check intake system hoses and renew if hard or cracked. Install turbocharger on exhaust manifold using new gasket. Attach remaining hoses, tubes and connections securely. Before starting engine, hold throttle to rear and crank engine with starter until oil light goes out. Start engine and check for oil or air leaks.

91. **TEST.** Radial and axial end play of turbine shaft can be used to indicate condition of thrust bearing, radial bearing and/or rotating parts. Carbon build-up can result in false readings, therefore accurate readings necessitate removal and cleaning. General condition can often be determined, however, by measuring radial and axial play before disassembling.

Refer to Fig. 90. Attach dial indicator to turbocharger mounting base with indicator rod inserted into return oil outlet in center housing and end contacting shaft. Move both ends of turbocharger shaft simultaneously alternately toward, then away from dial indicator. Radial play should be 0.08-0.13 mm new, with wear limit of 0.15 mm.

Refer to Fig. 91. To measure axial (end) play of turbocharger shaft, first unbolt and remove compressor housing (4 – Fig. 92). Mount dial indicator as shown in Fig. 91, with indicator rod resting on end of shaft. Move shaft back and forth and notice amount of movement recorded on dial indicator. Axial play should be 0.05-0.08 mm new, with wear limit of 0.13 mm.

92. **OVERHAUL.** Remove turbocharger unit as outlined in paragraph 90. Mark compressor housing (4 – Fig. 92), seal plate (16), center housing (25) and turbine housing (8) to aid alignment when assembling.

CAUTION: Do not rest weight of any parts of impeller on turbine blades. **Weight of only the turbocharger unit is enough to damage blades.**

If not already removed, remove screws (1), washers (2), clamp plate (3) and compressor housing (4). Check end play of turbine shaft as described in paragraph 91 to determine extent of wear before disassembling. Straighten tabs of lockplate (6), then remove nuts (5), lockplates (6), clamp plate (7) and turbine housing (8). It may be necessary to use a soft hammer to dislodge turbine housing from center housing. Be careful not to damage turbine wheel.

Clamp 11 mm box end wrench in vise, then position hex at center of turbine wheel (11) in wrench. Remove **left hand thread** nut (10) with 10 mm wrench. Remove compressor wheel (9) from shaft, then withdraw shaft (11) from center housing. Remove and discard old seal (12). Remove screws (14). New screws (14) should be used when assembling. Remove seal plate (16), then withdraw oil thrower (18) and seal rings (17). Remove screws (19) then withdraw thrust plate (21) and thrust bushing (22). Screws (19) should be new when reassembling. Remove snap rings (23) and bushings (24). New snap rings (23) should be used when assembling.

Bearing surfaces must all be smooth, especially check bores in center housing (25) and surface of turbine shaft (11).

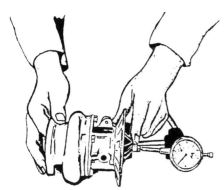

Fig. 90 — Dial indicator should be attached as shown for checking bearing radial clearance.

Fig. 91 — Dial indicator can be used to measure shaft end play.

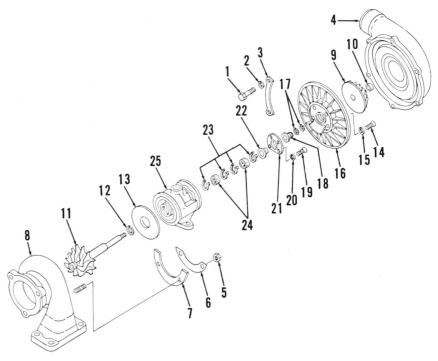

Fig. 92 — Exploded view of IHI turbocharger. Refer to test for disassembly procedure.

1. Screw	8. Turbine housing
2. Washer	9. Compressor wheel
3. Clamp plate	10. Nut (left hand thread)
4. Compressor housing	11. Turbine wheel and shaft
5. Nut	12. Oil seal
6. Lockplate	
7. Clamp plate	

13. Heat protector
14. Screw
15. Lockwasher
16. Seal plate
17. Seal rings
18. Oil thrower

19. Screw
20. Lockwasher
21. Thrust plate
22. Thrust bushing
23. Snap rings
24. Bushings
25. Center housing

Bore (A–Fig. 93) of center housing should not exceed 17.07 mm. Inside diameter (B) of center housing for bearings should not exceed 15.10 mm. Measure inside diameter and outside diameter of bearings (24–Fig. 92). If inside diameter exceeds 10.45 mm or outside diameter is less than 14.98 mm, bushings (24) must be renewed. Renew thrust plate (21) if thickness is less than 3.98 mm. Position thrust bushing (22) and oil thrower (18) on turbine shaft (11). Measure thrust bushing depth (C–Fig. 93) which should not exceed 4.09 mm. Width of grooves on oil thrower should not exceed 1.32 mm. Examine turbine blades carefully. Slight but even erosion is permissible and bearing journals can be polished lightly using crocus cloth. Diameter of shaft at bearing journals (E–Fig. 93) should not be less than 10.38 mm. Groove (F) for seal should not exceed 1.43 mm. With shaft supported in "V" block (V), deflection (G) should not exceed 0.011 mm. Check condition of compressor wheel (9–Fig. 92), compressor housing (4), heat protector (13) and turbine housing (8). Light rubbing can usually be cleaned-up and part reused, but renewal is necessary if cracked, warped or rubbed heavily.

Proceed as follows when assembling: Use new snap rings (23), screws (14 & 19), washers (15 & 20), lock plates (6), seals (12 & 17) and snap rings (23) when assembling. Install inner snap ring (23) with rounded shoulder toward bearing (24); lubricate and install bearing. Install second snap ring (23) with rounded shoulder toward installed bearing. Install third snap ring with rounded shoulder out, then lubricate and install second bearing (24). Install final snap ring with rounded side in toward bearing. Lubricate thrust bearing (22) and position bearing (22) and plate (21) in center housing. Install new washers (20) and screws (19) tightening screws to 0.8 N·m (7 in.-lbs.) torque. Use special installation tool (JDF-26-2) available from manufacturer for installing seals (17) on oil thrower (18). Install oil thrower and seals in seal plate (16). Apply thin coat of Dow Corning/RTV Silicon Sealant to surface of center housing, then install seal plate using new washers (15) and screws (14). Tighten screws (14) to 2.2 N·m (22 in.-lbs.) torque. Use special installation tool (JDF-26-1) available from manufacturer for installing seal (12) on turbine shaft (11). Install heat protector (13) on turbine side of center housing, then insert turbine shaft into center housing. A small tap on end of turbine shaft should seat shaft in place. Hold turbine shaft in place, turn shaft and housing over, install compressor wheel (9) and install nut (10). Nut (10) has left hand thread and should be tightened to

5.0 N·m (50 in.-lbs.) torque. Refer to paragraph 91 to check for clearances after assembling. Coat edge of seal plate (16) with Dow Corning/RTV Silicon Sealant and install compressor housing (4) with previously affixed marks aligned. Torque cap screws (1) to 3.7 N·m (37 in.-lbs.). Install turbine housing (8) with previously affixed marks aligned, tighten nuts (5) to 9.3 N·m (93 in.-lbs.) torque and lock by bending tabs of plate (6) around nuts.

Refer to paragraph 90 for installation procedures. Be sure turbocharger is full of oil before starting engine.

COOLING SYSTEM

TESTS

All Models

95. Cooling system should be operated

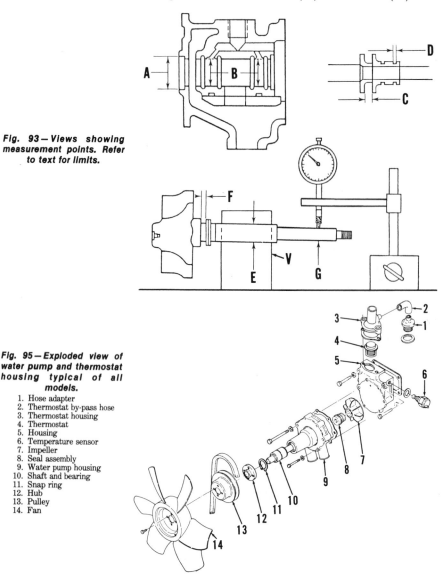

Fig. 93 – Views showing measurement points. Refer to text for limits.

at 0.89 bar (13 psi) for all models. Cooling system pressure is maintained by radiator cap. Capacity of cooling system is 6.6 liters (7 U S quarts).

Fan belt tension should be set to deflect 10-15 mm ($^3/_8$-$^5/_8$ in.) when pressed with 88.6 N(20 lbs.) force midway between water pump pulley and crankshaft pulley.

Thermostat (4–Fig. 95) is located under housing (3). Thermostat should open at 71 degrees C (160 degrees F.).

WATER PUMP

All Models

96. To remove water pump, drain cooling system, remove fan belt, disconnect coolant hoses from pump and unbolt radiator shroud. Unbolt water pump (9–Fig. 95) from housing (5), then withdraw pump and radiator shroud.

Press hub (12) from shaft (10). On

Fig. 95 – Exploded view of water pump and thermostat housing typical of all models.

1. Hose adapter
2. Thermostat by-pass hose
3. Thermostat housing
4. Thermostat
5. Housing
6. Temperature sensor
7. Impeller
8. Seal assembly
9. Water pump housing
10. Shaft and bearing
11. Snap ring
12. Hub
13. Pulley
14. Fan

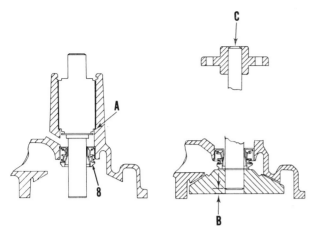

Fig. 96 – View showing cross-sections of water pump used on 850 and 950 models. Refer to Fig. 95 and text.

models so equipped, remove snap ring (11), then press bearing and shaft (10) forward out of housing (9) and impeller (7).

Press seal (8) into position in housing, then press bearing and shaft (10) into housing until bearing bottoms on shoulder (A – Fig. 96). Press fan and pulley hub onto shaft until flush (C – Fig. 96) on 850 and 950 models; or 8.5 mm past flush (C – Fig. 97) on 1050 models. Press impeller onto shaft using a spacer to support hub of impeller while pressing. Distance (B – Fig. 96) should be 4.8 mm for 850 and 950 models; distance (B – Fig. 97) should be 9.5 mm for 1050 models.

ELECTRICAL SYSTEM

ALTERNATOR AND REGULATOR

All Models

100. A Nippon Denso alternator is used on all models. The voltage regulator is located below the instrument panel and can be serviced after

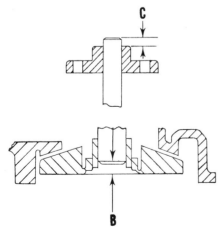

Fig. 97 – Cross-section showing setting dimensions (C & B) for assembling water pump on 1050 models. Refer to text.

removing the small access panel below key switch. To check alternator output, disconnect large wire from terminal "A" on rear of alternator and connect ammeter between terminal "A" and disconnected wire.

NOTE: Follow instructions for ammeter. Ammeter must be capable of reading at least 30 amps and some will require a shunt to be installed.

Detach connector from "N" and "F" terminals, then attach jumper wire with female connector to "F" terminal. Start engine and operate at 2000 rpm, touch jumper wire from "F" terminal to the "A" terminal and observe alternator output.

CAUTION: Check output quickly to prevent possible damage to alternator.

Alternator output should be 20 amps or more. If test indicates less than 20 amps, alternator is faulty. If test indicates 20 amps or more, regulator may be damaged.

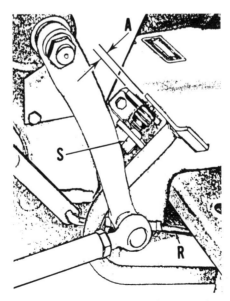

Fig. 100 – Starter safety switch (S) is located as shown on 850 and 950 models. Refer to text for adjustment.

Refer to the following data for recommended charging system specifications:
Alternator Brush Exposed Length-
  Minimum . . . . 5.5 mm
Alternator Pulley Nut Torque . . 54 N·m
Belt deflection with 89 N force applied midway between crankshaft and water pump pulleys should be 10-16 mm.

On 850 models before serial number 004739 and 950 models before serial number 007143, a wiring harness extension may be located between regulator and wiring harness. A diode (field discharge diode) is located in the extension harness to prevent "OIL" or "CHG" indicator lamps from burning out when switch is turned off. On later models, the field discharge diode is located on underside of voltage regulator and extension harness is not required.

STARTER

All Models

101. **SAFETY SWITCH.** On 850 and 950 models, a starting safety switch (S – Fig. 100) is located under clutch pedal pad and starting should be impossible unless clutch pedal is depressed. On 1050 models, a starting safety switch (S – Fig. 101) is located on right side of transmission and starting should be impossible unless range shift lever (R) is in neutral position.

Adjustment of safety switch for 850 and 950 models is accomplished as follows: Disconnect clutch rod (R – Fig. 100) from clutch pedal, then allow clutch pedal to rest on top of switch plunger without depressing plunger. Measure distance (A) between clutch pedal and stop plate. Distance (A) should be within limits of switch plunger movement (5.5-6.5 mm). If measured distance is not

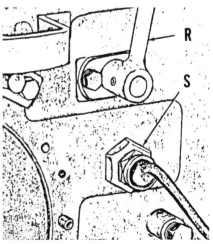

Fig. 101 – Starter safety switch (S) is located as shown. Range shift lever (R) should be in neutral to start engine.

correct, loosen locknut and rotate switch body to provide proper height.

102. **STARTER MOTOR AND DRIVE.** The starter motor and drive shown in lower view of Fig. 102 is type used on 850 models; type shown in upper view is typical of starter used on 950 and 1050 models. To remove either type of starter, disconnect battery ground cable, disconnect electrical connections to starter, then unbolt and remove complete starter and drive assembly by removing the two nuts. Refer to the following specification data:

850 Models

Brush Length-Minimum . . . . 9.3 mm
No Load Test-
    Volts 11.5
    Amperes . . . . .90 Max.
    Rpm . . . . .4000 Min.

950 & 1050 Models

Brush Length-Minimum . . . . . . 12.7 mm
No Load Test-
    Volts 11.0
    Amperes . . . . .180 Max.
    Rpm . . . . .3500 Min.

To remove drive, compress pinion (12 – Fig. 102) onto shaft (10) by clamping in vise. Use 15 mm (5/8 in.) I.D. tube to drive retainer (15) back, then remove snap ring (16).

ENGINE CLUTCH

ADJUSTMENT

All Models

105. Clutch pedal linkage is properly adjusted when pedal has free travel of 15-25 mm. Free travel is measured at pedal pad. To adjust, loosen locknuts on clutch rod (R – Fig. 100), then turn adjustment nut until free travel measured at pedal is within limit. Front adjustment locknut has left hand thread, rear locknut has standard thread.

On 850 and 950 models, check and adjust starter safety switch as outlined in paragraph 101.

On 1050 models, refer to paragraph 108 for adjustment of pto clutch release screws (F – Fig. 108).

CLUTCH UNIT

850 and 950 Models

106. **REMOVE AND REINSTALL.** The engine clutch can be removed from flywheel after first separating engine from clutch housing as outlined in paragraph 36.

Reverse removal procedure to reinstall clutch assembly. Long hub of

clutch lined disc should be toward rear. Clutch aligning tool (JDG-49 or equivalent) should be used to align clutch lined disc when assembling. Tighten cover to flywheel retaining screws to 27 N·m torque. Adjust clutch linkage as outlined in paragraph 105.

107. **OVERHAUL.** Clutch lined disc thickness (A – Fig. 106) should be 8.3-8.9 mm new. If thickness is less than 7.6 mm, install new lined disc. When compressed, new disc will be 7.8 mm thick.

Release lever setting is measured from machined facing surface of flywheel to rear surface of release lever plate, with a 7.7-7.9 mm gage block installed in place of friction disc. Gage block can be locally manufactured 240 mm outside diameter, 7.7-7.9 mm thick with 48 mm diameter hole in the center. A new clutch disc should have correct compressed thickness and may be used

to check release lever setting. To measure, install new clutch disc or gage block. Use depth gage to measure distance from release lever plate to edge of pilot bearing (B – Fig. 106). Remove clutch, place straightedge across friction surface of flywheel and measure distance (C) from rear edge of pilot bearing to straightedge (friction surface). Subtract measured distance "C" from measured distance "B". This difference should be 61-62.5 mm.

1050 Models

108. **REMOVE AND REINSTALL.** The engine clutch can be removed from flywheel after first separating engine from clutch housing as outlined in paragraph 36. Use special clutch aligning tool (JD-46) when removing, to keep pto clutch disc from falling. Be careful not to drop pto clutch disc.

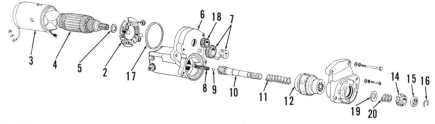

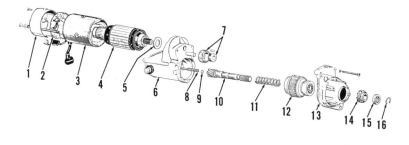

Fig. 102 – Exploded view of starter motor assemblies and drives used. Unit at top is used on 950 and 1050 models, lower unit is used on 850 models.

1. Cover	6. Reduction housing	11. Spring	16. Snap ring
2. Brushes and plate	7. Over-running clutch	12. Pinion	17. Seal
3. Housing and fields	8. Spring	13. Nose housing	18. Gear
4. Armature	9. Ball	14. Pinion	19. Washer
5. Thrust washer	10. Shaft	15. Retainer	20. Spring

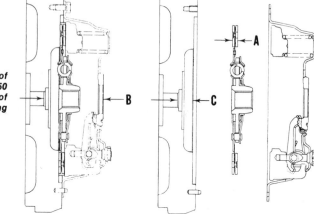

Fig. 106 – Cross-section of clutch used on 850 and 950 models showing points of measurement for adjusting release lever height.

Attach clutch assembly to flywheel using special aligning tool. Clutch retaining screws should be tightened to 22 N·m torque. Adjust pto clutch as follows:

Loosen locknuts (L–Fig. 108) and turn screw (S) until clearance between head of cap screw and plate is 1.5 mm and exactly the same for all three adjusting screws. Hold adjustment screw position and tighten locknut to 18 N·m torque, then recheck to be sure that 1.5 mm clearance has not changed.

Special adjusting gage (JDG-52) is available for setting release lever adjusting screws (F). All three screws must be exactly the same height. Gage measures finger height from flywheel surface which should be 112.3-113.7 mm. Be sure locknuts (N) are tight.

Complete reassembly by reversing removal procedure. Adjust clutch linkage as described in paragraph 105.

109. **OVERHAUL.** Before disassembling, punch mark the two parts of cover casting and the two pressure plates so that these four parts can be aligned the same when assembling. The pressure plates can be marked at one of the adjusting bolt locations and cover can be marked at one of holes for screws attaching clutch to flywheel. Clutch can be disassembled, by loosening locknuts (L–Fig. 108), then removing the three adjusting screws (S). All friction surfaces must be flat within 0.2 mm. Friction disc for transmission should be renewed if less than 7.6 mm. Renew pto friction disc if less than 7.0 mm.

When assembling, be sure to align marks that were installed before disassembly. Long hub (P) of transmission clutch disc must be toward rear. Refer to paragraph 108 for adjustment of screws (S) and release lever adjusting screws (F). Refer to paragraph 105 for adjustment of clutch linkage.

RELEASE BEARING

All Models

110. The release bearing can be removed and reinstalled after first separating engine from clutch housing as outlined in paragraph 36.

On 850 and 950 models, diameter of clutch shaft at release bearing should be 24.840-24.854 mm. Clearance between release sleeve bushing and clutch shaft should not exceed 0.5 mm. Inside diameter of sleeve bushing should be 24.96-24.97 mm. Inside diameter of release bearing should be 44.98-45.00 mm and outside diameter of sleeve bushing should be 45.00-45.02 mm.

On 1050 models, diameter of transmission clutch drive shaft where release bearing sleeve operates should be 34.70-34.80 mm. Clearance between sleeve bushing and transmission clutch shaft should not exceed 1.0 mm. Inside diameter of release sleeve bushings should be 34.99-35.05 mm. Free length of release bearing return springs should be 152 mm.

CLUTCH SHAFT

All Models

115. The clutch shaft can be withdrawn from front of 850 and 950 models after separating engine from clutch housing as outlined in paragraph 36.

The pto clutch shaft can be withdrawn from 1050 models after separating engine from clutch housing as outlined in paragraph 36; however, engine clutch shaft must be removed from rear. To remove the engine clutch shaft, separate engine from transmission and withdraw pto clutch shaft. Separate clutch housing from transmission as outlined in paragraph 127, then bump shaft with a soft hammer from front and withdraw from rear of housing.

On all models, refer to the following specification data:

850 and 950 Models
Clutch Shaft Diameter, at:
  Pilot Bearing
    Journal ....14.935-14.953 mm
  Release Bearing
    Sleeve ....24.840-24.854 mm
Release Sleeve Bushing to Clutch
Shaft Clearance-
  New.................0.11-0.13 mm
  Wear Limit ....0.5 mm
Clutch Shaft Pilot to Pilot Bushing,
Clearance-
  New.............0.0127-0.0508 mm
  Wear Limit ....0.3 mm
1050 Models
Pto Clutch Shaft Diameter, at:
  Pilot Bearing
    Journal ....16.98-16.99 mm

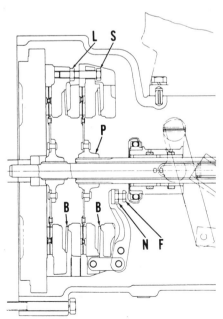

Fig. 108—Cross-section of dual stage clutch used on 1050 models. Refer to text.

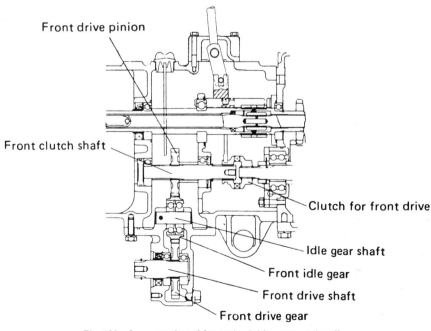

Fig. 120—Cross-section of front wheel drive power take off.

Transmission Clutch Shaft Diameter, at:
Release Bearing Sleeve
New . . . . .34.70-34.80 mm
Wear Limit . . . . .34.40 mm
Release Sleeve Bushing to Clutch
Shaft Clearance-
New . . . . .0.148-0.325 mm
Wear Limit . . . . .1.0 mm

FRONT DRIVE PTO

Power take off for driving the front differential is located in rear section of clutch housing.

R&R AND OVERHAUL

All Models So Equipped

120. The drop housing (24 – Fig. 121), gears (21 & 28) and related parts can be unbolted and removed after draining transmission lubricant from plug (25) and detaching drive shaft as follows:

Loosen drive shaft front cover clamp, remove screws attaching cover to drop housing, then slide cover toward front. Disconnect rear coupling by pushing shaft forward against spring coupling. Be careful not to lose steel balls from drive collars.

To remove gear (10), shaft (8) and related parts, refer to paragraph 115 and remove pto and transmission clutch shafts. Remove pin (15) and withdraw detent (14). Remove pin (18), then withdraw shaft (17) and shift fork (16). Remove drop housing (24) and snap ring (6) from rear bearing bore in clutch housing. Attach puller to threaded rear end of shaft (8) and pull shaft toward rear. When bearing (7) is free from bore, work through bottom opening from drop housing and remove snap ring (5) from front end of shaft. Continue pulling shaft out toward rear.

Idler shaft (22) is pressed into bores of housing (24) and held in place with pin (23). To remove, drive pin (23) out, then use 22 mm (⅞ in.) inside diameter pipe to receive shaft (22) while pressing from bores. To remove shaft (30), remove snap ring (29) from front, seal (33) and snap ring (32). Front of shaft (30) is threaded for puller which must be used to pull shaft (30) and bearing (31) from bores. Gear (28) will be free to fall from housing and bearing (27) can be pulled from housing if renewal is required.

Free length of detent spring (14) should be 33.8 mm and should exert 75 N when compressed to 26.4 mm. Forked end of fork (16) should be 6.7-6.9 mm thick and groove in coupler (4) should be 7.1-7.3 mm thick. Maximum wear limit for fork and coupler is when clearance

exceeds 1.0 mm. Clearance between shaft (17) and fork (16) should not exceed 0.5 mm. Diameter of shaft (17) should be 14.97-14.98 mm and inside diameter of bore in fork (16) should be 15.00-15.04 mm.

When assembling, spacer washer (20) should be installed on thickest side of gear (21) when measuring from ring in inner bore to outside edge. Use 25.4 mm pipe to drive bearing (7) into location, then install snap rings (5 & 6). Tighten drop housing retaining screws to 45-60 N·m torque.

TRANSMISSION

REMOVE AND REINSTALL

All Models

127. To detach (split) tractor between clutch housing and front of transmission, proceed as follows: Disconnect battery ground, then remove tractor seat, brake actuating rods and brake return springs. Unbolt and remove tractor step plates from both sides. Detach hydraulic lines from right side and power steering pressure and return lines from models so equipped.

On 1050 models, remove front wheel

drive cover plate (if equipped with four wheel drive) or cover plate, then use screwdriver to wedge against clutch shaft to be sure that clutch shaft stays in place when separating. If clutch shaft is permitted to move rearward while separating, a thrust washer may fall or a clutch lined disc may move requiring separation of clutch housing from engine to reassemble.

On all models, support both front and rear sections of tractor, then remove screws which attach clutch housing to transmission housing. Carefully separate front and rear sections of tractor. Refer to paragraph 150 for removal of axle housings if removal is necessary.

When assembling, tighten clutch housing to transmission housing attaching screws to 120 N·m torque.

SHIFT LEVER AND FORKS

All Models

128. The shift lever (1 – Fig. 128) and top cover (11) can be removed without removing other parts; however, removal of shift rails (18, 19 and 20) and forks (23, 24 and 25) requires separation of transmission from clutch housing as outlined in paragraph 127.

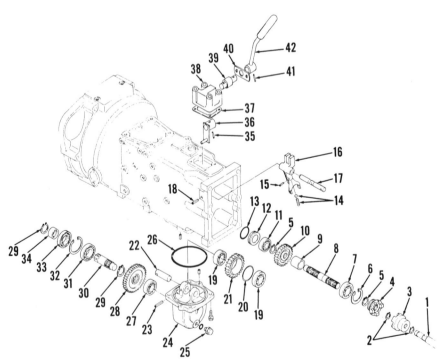

Fig. 121 – Exploded view of front wheel drive power take off.

1. Input shaft	12. Plug	23. Pin
2. Snap rings	13. "O" ring	24. Drop housing
3. Coupler	14. Detent spring and ball	25. Plug
4. Coupler	15. Pin	26. "O" ring
5. Snap ring	16. Shift fork	27. Bearing
6. Snap ring	17. Shaft	28. Front drive gear
7. Bearing	18. Pin	29. Snap ring
8. Front clutch shaft	19. Bearings	30. Shaft
9. Spacer	20. Spacer washer	31. Bearing
10. Front drive pinion	21. Idler gear	32. Snap ring
11. Bearing	22. Idler shaft	33. Seal

34. Shield
35. Pin
36. Shift arm
37. Gasket
38. Cover
39. Shift rod
40. Retainer plate
41. Pin
42. Four wheel drive engagement lever

Remove dip stick (10) and shift transmission to neutral before unbolting shift cover (11) from transmission housing. Be careful and do not drop detent springs (14) into transmission housing while removing shift cover.

Remove detent springs and balls (14) from three holes in top of housing at rear of shift rails. Separate transmission housing from clutch housing as outlined in paragraph 127. Unbolt and remove differential lock cover (12), reverse stop plate (13) and, on 850 and 950 models, retainer plate (16). Bend a length of wire into "J" shape so that wire can be held from above and will curve up and enter roll pins (22) in shift rails from the bottom. Drive the roll pins down, catching them with wire to prevent pins from falling into transmission housing. Drive shift rail (18) forward out of housing. End seals (17), used on 850 and 950 models, will be removed as shafts are driven forward. On all models, lift shift fork (23) out top as rail (18) is removed. Proceed by removing center rail (19) and fork (24), then finally removing rail (20) and fork (25). Rail (20) has additional roll pin which must be removed. Interlock pins (21) can be removed through hole for plug (26).

Inspect end seals (17), on 850 and 950 models, and renew if leaking or otherwise damaged. On all models, renew detent springs and balls (14) if damaged. Free length of springs should be 33.8 mm and springs should exert 72-79 Newtons when compressed to working height of 26.4 mm. Fork ends of shift forks (23, 24 and 25) should be 6.7-6.9 mm thick and groove in shift collars should be 7.1-7.3 mm. Renew fork and/or shift collar if clearance exceeds 1.0 mm. Shift rails (18, 19 and 20) should be 14.96-14.98 mm diameter and bore in shift forks should be 15.00-15.04 mm. If clearance exceeds 0.2 mm, renew rail or fork.

To reinstall, position fork (23) in groove of sliding gear collar, then install shift rail (18). Shift rails (18 and 19) are identical. Install roll pin through shift fork (23) and rail (18) with split in pin toward front of tractor. Install shift rails (19 and 20) and forks (24 and 25) similarly, making sure that pins (21) are properly located between rails. Roll pins (22) should be approximately 0.5 mm below flush with forks. The stop pin at rear of reverse rail (20) should protrude approximately 10 mm from shaft. Install stopper plate (13), with screws torqued to 27 N·m. Angle cut corners of plate (13) must be toward rear. Check to be sure that engagement is correct and that interlock will only permit one rail to be moved from neutral position at a time, then install plug (26). Install end seals (17) and retainer plate (16) on 850 and

950 models. Tighten retainer plate nuts to 11 N·m torque. On all models, pin (4) is 40 mm long and should protrude evenly from both sides of ball (5). Tighten screws retaining covers (6 and 11) to 27 N·m torque.

REVERSE IDLER

All Models

129. The reverse idler gear (32 – Fig. 128) and shaft (29) can be removed after separating clutch housing from transmission as outlined in paragraph 127. Only approximately 4 liters of transmission lubricant needs to be drained from transmission. Remove cover (15) from left side, shift transmission to fourth gear. Bend a wire that will fit inside roll pin into "J" shape and insert into roll pin from back side. Hold end of wire in large opening, drive roll pin into transmission housing. Catch pin with wire and remove, then use punch in

roll pin hole to pry shaft (29) forward. Carefully catch reverse gear as shaft is withdrawn.

Inside diameter of bushings (31) should be 20.02-20.09 mm and shaft (29) should have outside diameter of 19.99-20.00 mm. Clearance between shaft and bushings must not exceed 0.2 mm. Bushings should be flush with ends. Lubricate "O" ring (28) to ease entry into case bore. Shift collar in gear should be toward rear and should engage shift fork if installed. Align holes in shaft and housing for 36 mm long roll pin (30). Retaining nut and screws for cover (15) should be tightened to 27 N·m torque.

TOP (DRIVE) SHAFT

850 and 950 Models

130. Remove the shifter cover, rails and forks as outlined in paragraph 128. If differential drive shaft needs to be re-

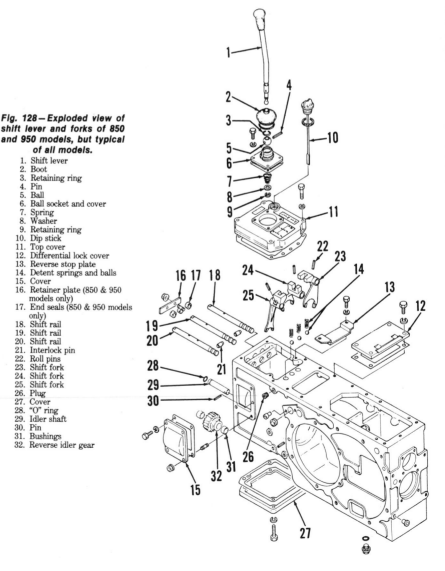

Fig. 128 – Exploded view of shift lever and forks of 850 and 950 models, but typical of all models.

1. Shift lever
2. Boot
3. Retaining ring
4. Pin
5. Ball
6. Ball socket and cover
7. Spring
8. Washer
9. Retaining ring
10. Dip stick
11. Top cover
12. Differential lock cover
13. Reverse stop plate
14. Detent springs and balls
15. Cover
16. Retainer plate (850 & 950 models only)
17. End seals (850 & 950 models only)
18. Shift rail
19. Shift rail
20. Shift rail
21. Interlock pin
22. Roll pins
23. Shift fork
24. Shift fork
25. Shift fork
26. Plug
27. Cover
28. "O" ring
29. Idler shaft
30. Pin
31. Bushings
32. Reverse idler gear

moved, unbolt and remove cover (53 – Fig. 130), then loosen nut (59). Be careful not to lose or damage shims (57 and 58).

Remove cover (33) and attach suitable puller (JDT-40 and JDE-114-1 or eqivalent) to threaded end of shaft (42). Position suitable spacer (JDT-39 or equivalent) between gears (40 and 41). Position 1⅝-inch open end wrench (or similarly constructed ½-inch thick spacer) between gears (41 and 48). Work through opening usually covered by cover (12) and remove snap ring (52). Pull shaft (42) forward approximately 6 mm, withdraw 1⅝-inch open end wrench (or spacer), dislodge snap ring (44) and move snap ring toward rear of shaft. Reposition 1⅝-inch open end wrench (or spacer) and continue pulling shaft out toward front. Remove parts as shaft moves forward. Splined coupling (between transmission top shaft and pto drive shaft), snap ring (52) and washer (51) must be withdrawn through opening normally covered by cover (12). Remove thrust washer (49), continue pulling shaft forward, then withdraw low reduction gears (48), bearings (47), snap rings (46), thrust washer (45) and snap ring (44). Remainder of parts and shaft can be withdrawn from case. Rear bearing (50) must be driven from bore toward rear.

Differential drive shaft and gear must be installed before top shaft. To assure correct assembly of top shaft rear bearing thrust washer (49) over pin (43), bearing (50) must be installed from rear. To install bearing from rear, it is necessary to remove the pto drive shaft as outlined in paragraph 161.

Install washer (39) against shoulder on front of shaft. Bearing (38) must have retaining ring on outside diameter toward front of shaft and snap ring (37) must be completely seated in groove. Use grease to hold pin (43) in hole in shaft. Position low reduction gear (48) in case. Be sure that needle bearings (47) stay in gear. Start shaft (42) through front opening, positioning first and second sliding gear (40) over end of shaft, then start third and fourth sliding gear (41) over shaft. Continue sliding shaft into gears until snap ring (44) and washer (45) can be started on shaft. Flat side of thrust washer must be toward front, tanged inside diameter locates washer on splines. Continue sliding shaft through gear (48) being careful not to damage needle bearings (47). Seat snap ring (44) in groove. Grease thrust washer (49) and install over pin (43) with chamfered side toward front. Insert a feeler gage between top of transmission case and in front of rear wall for bearing to hold thrust washer (49) against the low reduction gear (48).

Drive bearing (50) into location in case bore and on shaft. Remove feeler gage used to hold thrust washer temporarily on shaft pin, then finish driving bearing into position until snap ring groove is visible behind bearing. Install washer (51) and snap ring (52), being sure that snap ring enters groove. Check sliding gears for operation and install retainer (33). Install pto shaft as outlined in paragraph 161 and shift rails and forks as outlined in paragraph 128. Screws attaching retainer (33) should be tightened to 27 N·m torque.

1050 Models

131. Remove the shifter cover, rails and forks as outlined in paragraph 128. If differential drive shaft needs to be removed, remove snap ring (94 – Fig. 131) and oil slinger (93) or drive coupling (from four wheel drive models). Straighten lock plate (60) then loosen nut (59).

Remove retainer (33) and withdraw pto drive shaft. Position a 58.8 cm long section of threaded rod through shaft (42) with 41 mm OD washer and nut at rear end and JDT-40 or equivalent puller at front. Position suitable spacer (JDT-39 or equivalent) between gears (40 and 41). Position 1⅝-inch open end wrench (or similarly constructed ½-inch thick spacer) between gears (41 and 48). Work through opening usually covered by cover (12) and remove snap ring (52)

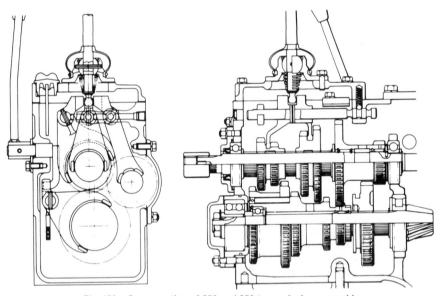

Fig. 129 – Cross-section of 850 and 950 transmission assembly.

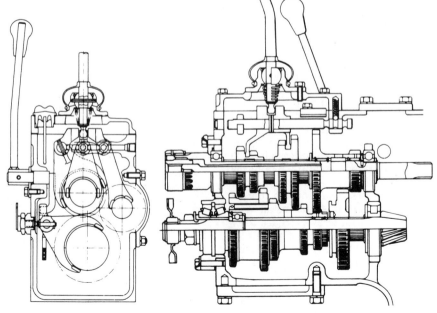

Fig. 129A – Cross-section of 1050 two wheel drive transmission assembly.

and washer (51). Pull shaft (42) forward approximately 6 mm, withdraw 1⅝-inch open end wrench (or spacer), dislodge snap ring (44) and move snap ring toward rear of shaft. Reposition 1⅝-inch open end wrench (or spacer) and continue pulling shaft out toward front. Remove parts as shaft moves forward. Remove thrust washer (49), continue pulling shaft forward, then withdraw low reduction gears (48), bearings (47), snap rings (46), thrust washer (45) and snap ring (44). Remainder of parts and shaft can be withdrawn from case. Rear bearing (50) must be driven from bore toward rear.

Differential drive shaft and gear must be installed before top shaft. Inside diameter of pto shaft front bushing (90) should be 32.00-32.08 mm. Bushing should be at bottom of front chamfer in shaft. Inside diameter of rear pto bushing in shaft (42) should be 20.00-20.08 mm. Bushing should be 3 mm forward of chamfer in rear of shaft.

Press bearing (38) onto shaft (42) against shoulder of shaft with retaining ring on outside diameter toward front. Seat snap ring (37) completely in groove. Position low reduction gear (48) in case. Be sure that needle bearings (47) stay in

gear. Start shaft (42) through front opening, positioning first and second sliding gear (40) over end of shaft, then start third and fourth sliding gear (41) over shaft. Continue sliding shaft into gears until snap ring (44) and washer (45) can be started on shaft. Flat side of thrust washer must be toward front, tanged inside diameter locates washer on splines. Continue sliding shaft through gear (48) being careful not to damage needle bearings (47). Seat snap ring (44) in groove. Grease thrust washer (49) and install with chamfered side toward front. Insert a feeler gage between top of transmission case and in front of rear wall for bearing to hold thrust washer (49) against the low reduction gear (48). Drive bearing (50) into location in case bore and on shaft. Remove feeler gage used to hold thrust washer temporarily on shaft pin, then finish driving bearing into position until snap ring groove is visible behind bearing. Install washer (51) and snap ring (52), being sure that snap ring enters groove. Check sliding gears for operation, then install retainer (33) and shift rails and forks as outlined in paragraph 128. Screws attaching retainer (33) should be tightened to 27 N·m torque.

DIFFERENTIAL DRIVE SHAFT

850 and 950 Models

132. Remove the shifter cover, rails and forks as outlined in paragraph 128. Unbolt and remove cover (53 – Fig. 130), then loosen nut (59). Be careful not to lose or damage shims (57 and 58). Refer to paragraph 130 and complete removal of top (drive) shaft and gears.

Refer to paragraph 134 if the range shifter or hi-lo sliding gear must be removed; however, these parts do not need to be removed for service to differential drive shaft.

Refer to paragraph 150 and remove both final drive units. Refer to paragraph 140 to remove differential lock and paragraph 141 for removal of differential.

Install jack screws in two threaded holes of bearing retainer (55 – Fig. 130), then tighten screws enough to remove shims (57 and 58). Shims (57 and 58) adjust pinion position and will not need to be changed unless pinion and ring gear assembly is renewed or if adjustment is otherwise incorrect. Use a brass drift and bump differential shaft out toward rear. If not removed, align reverse idler

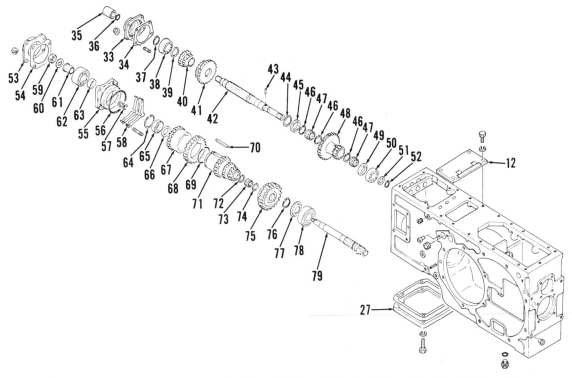

Fig. 130 – Exploded view of top drive shaft, differential drive shaft and related parts for 850 and 950 models.

12. Differential lock cover	40. First and second sliding gear	47. Bearings	56. "O" ring	65. Bearing	73. Needle bearing
27. Cover	41. Third and fourth sliding gear	48. Low reduction gears	57. Shims	66. Spacer	74. Snap ring
33. Retainer		49. Thrust washer	58. Shims	67. Gear	75. Hi-lo sliding gear
34. Gasket	42. Top shaft	50. Bearing	59. Nut	68. Gear	76. Snap ring
35. Coupling	43. Pin	51. Washer	60. Lockwasher	69. Retaining ring	77. Thrust washer
36. Snap ring	44. Snap ring	52. Snap ring	61. Bearing sleeve	70. Drive key	78. Roller bearing
37. Snap ring	45. Thrust washer	53. Cover	62. Bearing	71. Cluster gears and shaft	79. Differential drive shaft
38. Bearing	46. Snap rings	54. Gasket	63. Spacer	72. Snap ring	
39. Washer		55. Retainer	64. Snap ring		

gear teeth with front cluster gear to prevent binding. Shift hi-lo shifter to "Lo". Withdraw parts as they are removed from shaft.

Inspect all gear teeth and bearings carefully for wear or damage. Inside diameter of bearing sleeve (61) is 22.99-23.01 mm and outside diameter is 29.99-30.01 mm. Flat side of gear (68) should be toward gear (67).

To assemble, press inner race and roller of bearing (78) onto shaft (79) with flared end of roller cage toward bevel pinion. Install outer race of bearing (78) with shoulder toward front. Install thrust washer (77) with flat side toward front (away from bearing), then install snap ring (76) in groove. Position cluster assembly (64 through 74) in case with large end forward and hi-lo sliding gear (75) in case with fork collar end forward toward cluster gear. Insert differential shaft through differential compartment bore, hi-lo sliding gear (75) and cluster gear. Bump shaft forward through front bearing of cluster gear. Check to make certain that gear (75) moves freely on splines and to be sure that cluster gear rotates freely on bearings. Lubricate "O" ring (56) and install in groove of bearing housing (55), then position bearing housing in case. Install spacer (63) over end of shaft and press sleeve (61) into inner race of bearing (62). Position bearing and sleeve over end of shaft, brace rear

of differential drive shaft and bump bearing and sleeve onto front of shaft. Install lockwasher (60) and nut (59). Tighten nut (59) to 80-100 N·m torque, then lock by bending part of washer (60) around nut.

If bevel gears were not changed and original setting is believed to be satisfactory, install original shims (57 and 58). Be sure that shims (57) are identical in thickness to shims (58). If adjustment of bevel gear mesh position is questioned, refer to paragraph 145 for adjustment procedure. Install gasket (54) and cover (53) then tighten retaining nuts to 27 N·m torque.

Complete reassembly, by installing differential lock as in paragraph 140 and differential as outlined in paragraph 141, range shifter assembly as outlined in paragraph 134, transmission drive shaft as outlined in paragraph 130, pto drive shaft assembly as outlined in paragraph 161, and shift rails, forks and cover as outlined in paragraph 128.

1050 Models

133. Remove the shifter cover, rails and forks as outlined in paragraph 128. Remove snap ring (94 – Fig. 131) and oil slinger (93) or drive coupling (from four wheel drive models). Straighten lock plate (60), then loosen nut (59). Be

careful not to lose or damage shims (57 and 58). Refer to paragraph 131 and complete removal of top (drive) shaft and gears.

Refer to paragraph 134 if the range shifter or hi-lo sliding gear must be removed; however, these parts do not need to be removed for service to differential drive shaft.

Refer to paragraph 150 and remove both final drive units. Refer to paragraph 140 to remove differential lock and paragraph 141 to remove differential.

Unbolt bearing retainer (55 – Fig. 131), remove shims, then install jack screws in two threaded holes of bearing retainer. Tighten jack screws enough to remove shims (57 and 58). Shims (57 and 58) adjust pinion position and will not need to be changed unless pinion and ring gear assembly is renewed or if adjustment is otherwise incorrect. Use a brass drift and bump differential shaft out toward rear. If not removed, align reverse idler gear teeth with front cluster gear to prevent binding. Shift hi-lo shifter to "Lo". Withdraw parts as they are removed from shaft.

Inspect all gear teeth and bearings carefully for wear or damage. Inside diameter of bearing sleeve (61) is 23.99-24.01 mm and outside diameter is 29.98-30.01 mm.

To assemble, press inner race and

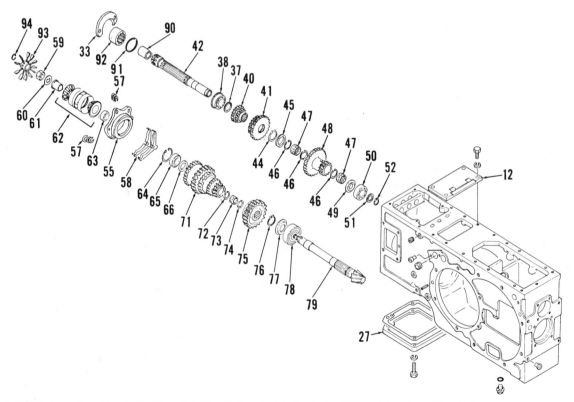

Fig. 131—Exploded view of top drive shaft, differential drive shaft and related parts for 1050 models refer to Fig. 130 for legend except the following.

90. Bushing	91. Snap ring	92. Coupling	93. Oil slinger	94. Snap ring

roller of bearing (78) onto shaft (79) with flared end of roller cage toward bevel pinion. Install outer race of bearing (78) with shoulder toward front. Install thrust washer (77) with flat side toward front (away from bearing), then install snap ring (76) in groove. Position cluster assembly (71) in case with large end forward and hi-lo sliding gear (75) in case with fork collar end forward toward cluster gear. Insert differential shaft through differential compartment bore, hi-lo sliding gear (75) and cluster gear. Bump shaft forward through front bearing of cluster gear. Check to make certain that gear (75) moves freely on splines and to be sure that cluster gear rotates freely on bearings. Press cups of bearings (62) against large snap ring installed in bearing housing (55). Press front bearing cone onto spacer (61), then insert cone and spacer into position in housing and bearing cup. Position small inner spacer over spacer (61) then press spacer (63) over end of shaft, then position assembled bearing (62) and rear bearing cone onto spacer until rear of bearing is flush with spacer. Bearing should be tight but rotate freely. Install sleeve over end of shaft, brace rear of differential drive shaft and bump bearing and sleeve onto front of shaft. Install

lockwasher (60) and nut (59). Tighten nut (59) to 200-250 N·m torque, then lock by bending part of washer (60) around nut.

If bevel gears were not changed and original setting is believed to be satisfactory, install original shims (57 and 58). Be sure that shims (57) are identical in thickness to shims (58). If adjustment of bevel gear mesh position is questioned, refer to paragraph 145 for adjustment procedure. Tighten nuts retaining housing (55) to 27 N·m torque.

Complete reassembly, by installing differential lock as in paragraph 140 and differential as outlined in paragraph 141, range shifter assembly as outlined in paragraph 134, transmission drive shaft as outlined in paragraph 131 and shift rails, forks and cover as outlined in paragraph 128.

RANGE SHIFTER ASSEMBLY

All Models

134. The range shifter assembly should only be removed if shift parts are damaged and must be removed or if the hi-lo sliding gear (75 – Fig. 130 or Fig. 131) must be renewed. Differential pinion shaft and other parts can be removed and reinstalled with range gear and shift parts in place. Installation of detent assembly (83 – Fig. 132) is very difficult and necessitates removal of final drive assembly.

To remove, first refer to paragraph 130 or 131 and remove transmission top shaft and paragraph 150 to remove right final drive. Remove pin (88 – Fig. 132) and lever (89). Unbolt and remove plate (87) and "O" ring (86), then withdraw arm shaft (85). Slide gear (75) forward, then drive pin (82) down and catch from falling using "J" shaped wire inserted in pin from bottom. Drive shaft (84) forward out of fork (80) and case bore. As shaft moves forward, detent ball and spring (83) will fall free. Ball may fall either into differential compartment or transmission compartment. Refer to paragraph 131 for removal and installation of differential drive shaft if gear (75) is to be renewed.

End of shift fork (80) should be 6.7-6.9 mm thick and should have maximum of 1 mm clearance in 7.1-7.3 mm collar groove. Inside diameter of fork is 14.96-14.98 mm and shaft (84) is 15.00-15.04 mm diameter. Detent spring (83) free length new is 33.8 mm and should exert 72-79 N when compressed to working length of 26.4 mm. Clearance between 19.95-20.00 mm shaft (85) and bore in case should not exceed 0.5 mm.

To assemble, start pin (82) in top of fork with opening in pin toward front.

Position fork in groove of gear (75) which is correctly positioned on bevel pinion shaft. Coat shaft (84) with oil then insert through front of transmission housing and into shift fork. Insert detent spring through hole (D – Fig. 133), use grease to stick 5/16-inch diameter detent ball onto end of ¼-inch diameter tubing and insert ball (and tubing) into hole (D). Compress detent spring with ball and tubing, push shaft (84) back over detent ball, then withdraw tubing. Be sure that shaft is back far enough to engage detent ball. Be sure hole in shaft for pin (82) is aligned with pin, then install pin through shaft. Install seal cap (81) and shifter arm (85). Coat "O" ring (86) with oil, position "O" ring over shaft, then install plate (87). Install shift lever (89) and pin (88).

Remainder of assembly is reversal of disassembly procedure.

DIFFERENTIAL, FINAL DRIVE AND REAR AXLE

DIFFERENTIAL LOCK SHIFTER

All Models

140. **REMOVE AND REINSTALL.** To remove the differential lock, unbolt bracket (3 – Fig. 140) and pedal (1) from right side of transmission housing and cover (12) from top. Bend a length of wire in "J" shape and insert from underside of roll pin (7), then drive pin down out of fork (8), catching pin with wire. Unbolt plate (4), then withdraw shaft (6) from side of transmission housing and

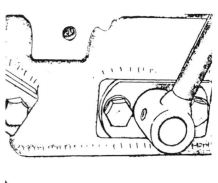

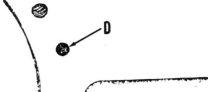

Fig. 133 – Refer to text for procedure for installing range shift detent (83 – Fig. 132) in hole (D) shown above.

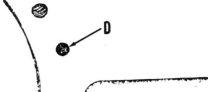

Fig. 132 – Exploded view of range shift assembly. Refer also to Fig. 130 for parts (75 through 79).

80. Range shift fork	85. Shift arm
81. Plug	86. "O" ring
82. Pin	87. Retainer
83. Detent ball and spring	88. Pin
84. Shaft	89. Range shift lever

spring (9) from top opening. Rockshaft housing must be removed to withdraw fork (8).

Free length of spring (9) is 96.5 mm and spring should exert 412 N when compressed to 64.5 mm.

When assembling, position spring (9) between fork (8) and left side of transmission housing. Special tool JDT-36 (JDT – Fig. 140) is available for compressing spring (9) so that shift fork can be positioned beside spring and shaft (6) can be inserted into fork before releasing spring. Install and lubricate "O" ring (5) in shaft groove. Install pin (7) through fork (8) and shaft (6). Screws attaching plate (4) and bracket (3) should be tightened to 27 N·m torque; screws attaching cover (12) should be tightened to 45 N·m torque.

DIFFERENTIAL AND BEVEL GEARS

All Models

141. R&R AND OVERHAUL. To remove the differential assembly, first remove rockshaft housing from top of center housing. Remove differential shifter as outlined in paragraph 140 and both final drive assemblies as outlined in paragraph 150. Remove screws from each of the retainer plates (26 – Fig. 140); then, drive differential out toward left side. Be careful to catch differential as differential carrier bearings are released from bores.

Ring gear (14) and bevel pinion are available only as a matched set. Refer to paragraph 132 or 133 for disassembly of

differential drive shaft and to paragraph 145 for adjustment of mesh position. Inside diameter of bushing (15) in right side should be 33.000-33.075 mm; inside diameter of bushing (31) in left side of differential should be 42.000-42.085 mm. Maximum clearance between pinions (18) and shaft (30) is 0.4 mm. New washers (17 & 22) are 0.95-1.05 mm thick. Install new thrust washer if less than 0.6 mm thick.

Grease thrust washers (17 & 22) and position on respective gears (18, 20 and 21). Assemble side gear (20) with holes for lock pins (13) in right side, and plain side gear (21) on left side of differential carrier (16). Assemble gears (18) and washers (17), turn side gears to align both gears (18) with holes for shaft (30), then install shaft (30). Install roll pin (19) through housing (16) and shaft (30) with split in pin positioned 90 degrees from shaft centerline. Heat bearings (11 and 23) and ring gear (14) to 150 degrees C. before assembling. Be sure that identification number etched on end of drive shaft matches number etched on ring gear. Install ring gear using three equally spaced screws to pull ring gear tight against carrier (16), then remove the three screws. Install lock plates (24), then install all six of the ring gear retaining screws evenly to 54 N·m torque. Bend ends of lock plates around screws to lock screws. Install heated left bearing (23) against step on housing. Install lock collar (13), then install one snap ring (10) in inner groove. Check lock collar (13) for freedom of movement. Install heated right side bearing (11), then install the outside snap ring (10).

If bevel ring gear or bevel pinion are renewed or if mesh (cone) position is questioned, refer to paragraph 145 for checking and adjustment.

Install differential assembly using shims (27) that were originally installed. Screws attaching retainer (28) and screws attaching plates (26) should both be tightened to 27 N·m torque. Fabricate and use a retaining strap described in Fig. 141 to hold front of side cover (25 – Fig. 140) flush with side of transmission housing. Measure gear backlash using a dial indicator at outer edge of ring gear teeth. Base of dial indicator should be attached to top of transmission housing. Correct backlash is 0.13-0.18 mm. Be sure that carrier (16) and bearing (23) are fully against retainer (28) when checking.

If backlash is excessive, be sure that retainer plates (26) and fabricated plate (Fig. 141) are tight and that side cover (25 – Fig. 140) is flush with machined surface of transmission housing. Removing some shims (27) will decrease backlash.

If backlash is less than 0.13 mm, be

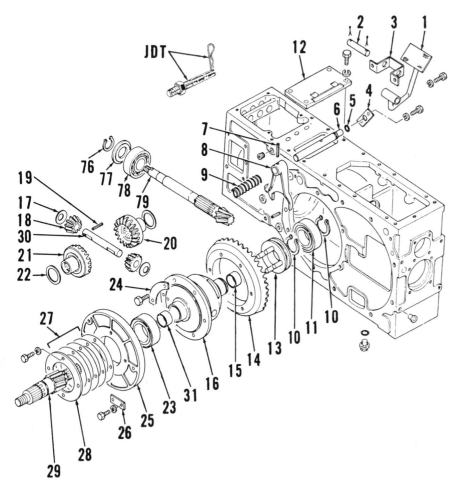

Fig. 140 — Exploded view of differential and final drive pinion typical of all models. Special tool (JDT) is used for compressing spring (9) so that shift fork (8) can be installed.

1. Differential lock pedal	19. Pin	28. Retainer
2. Pivot shaft	20. Side gear (right side)	29. Final drive pinion
3. Bracket	11. Carrier bearing	21. Side gear (left side)	30. Pinion shaft
4. Retainer	12. Differential lock cover	22. Thrust washers	31. Bushing
5. "O" ring	13. Locking collar	23. Carrier bearing	76. Snap ring
6. Shaft	14. Bevel ring gear	24. Lock plate	77. Thrust washer
7. Pin	15. Bushing	25. Side cover	78. Bearing
8. Fork	16. Differential housing	26. Retainer plate	79. Differential drive
9. Spring	17. Thrust bearings	27. Shims	shaft
10. Snap rings	18. Differential pinion gears		

sure that bearing (23) is tight against retainer (28) and that shoulder of carrier (16) is tight against bearing (23). Rotate gear (14) to three locations and recheck backlash. If all three backlash readings are not alike, be sure that ring gear is completely seated against carrier (16). Add shims at (27), then push carrier (16) and bearing (23) tight against retainer (28) before rechecking.

Refer to paragraph 150 for installation of final drive assemblies and to paragraph 140 for reinstalling differential shifter.

CONE (MESH) POSITION

All Models

145. Cone (mesh) position is adjusted by changing thickness of shims (57 & 58 – Fig. 144 or Fig. 145). Checking cone position necessitates special gage (S – Fig. 144) on 850 and 950 models and special GO/NO-GO gage (G – Fig. 145) for 1050 models. On all models, the transmission must be separated from clutch housing as outlined in 127 in order to change shims (57 & 58 – Fig. 144 or 145). Rockshaft housing must be removed to measure cone point.

On 850 and 950 models, install special (JDT–41) gage (S – Fig. 144) in place of differential, then measure distance between end of pinion and special tool with feeler gage (F). Clearance measured by feeler gage should be 0.45-0.55 mm. If incorrect, change thickness of shims (57 & 58) as necessary to correct clearance. Shims may be added or removed without removing pinion shaft. Remove screws and nut attaching cover (53) and bearing retainer (55), then install jack screws in two threaded holes provided to move bearing retainer (55) enough to add or remove shims (57 and 58). Be sure that thickness of shims (58) is exactly the same as thickness of shim (57).

On 1050 models, install differential assembly, then use special (JD-48-1) GO/NO-GO gage (G – Fig. 145) to measure clearance between differential carrier housing (16) and end of drive pi-

nion. Thin section of gage (G) should fit but thicker end should not fit. If incorrect, change thickness of shims (57 and 58) as necessary to correct clearance. Shims may be added or removed after loosening screws attaching bearing housing (55). Use jack screws in threaded holes of retainer (55) to move retainer forward enough to add or remove shims (57 and 58). Be sure that thickness of shims are exactly the same at each stack (57) and (58).

FINAL DRIVE

All Models

150. **REMOVE AND REINSTALL.** To remove the final drive assembly, drain oil from transmission housing, support tractor and final drive assembly, then remove rear wheel. Remove fender assembly, then remove all of the screws which attach axle housing to transmission/differential housing.

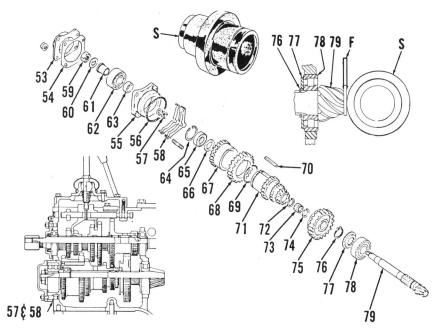

Fig. 144 — Views showing special tool (S) used to check mesh position of differential drive shaft pinion on 850 and 950 models. Refer to text.

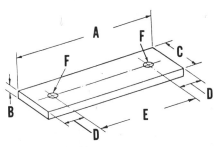

Fig. 141 — Drawing of plate used to retain spacer (25 – Fig. 140) while checking backlash and mesh position of gears (14 & 79 – Fig. 140).

A = 170 mm   D = 25 mm
B = 8 mm   E = 120 mm
C = 40 mm   F = 11 mm

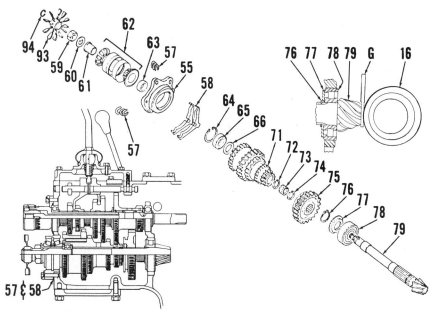

Fig. 145 — Views showing special gage (G) used to check mesh position of differential drive shaft pinion on 1050 models. Refer to text.

Use the longer retaining screws as jack screws in threaded holes provided to force axle housing off dowel pins and to separate gasket which has sealer applied to both sides.

When reinstalling, tighten axle housing to transmission/differential housing screws to 54 N·m torque.

151. OVERHAUL. The final drive must be removed as outlined in paragraph 150 in order to repair any part of the final drive. Remove brake

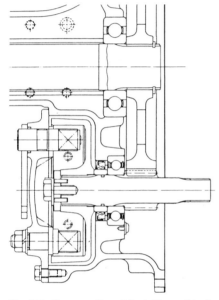

Fig. 150 — Cross-section of final drive and brake typical of 850 models.

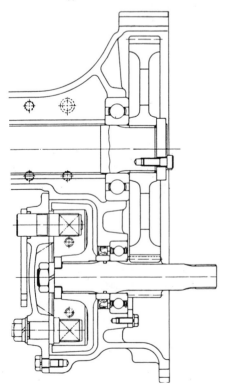

Fig. 151 — Cross-section of final drive and brake typical of 950 and 1050 models.

assembly as outlined in paragraph 155.

On 850 models, remove snap ring (16 – Fig. 152), then remove final drive gear (20). Remove and save any shims that may be located between gear (20) and bearing (21). Axle shaft (32) may be pressed from bearings (21 & 24), or retainer (28) may be unbolted and shaft pressed out of bearing (21), leaving bearing (24) on shaft. To remove pinion (1), remove the two Phillips head screws retaining bearing (4) in housing, then remove the brake drum and the final drive gear (20), if not already removed. Reinstall screw (14) and bump pinion from brake side toward differential until free from housing.

On 950 and 1050 models, remove screws (17), lock plate (18) and spacer (19), then lift final drive gear (20) from shaft. Axle shaft (32) may be pressed from bearings (21 & 24) or retainer (28) may be unbolted and shaft pressed out of bearing (21), leaving outer bearing (24) and seal on shaft. To remove pinion (1), unbolt retainer (2) from housing, then remove brake drum and final drive gear (20), if not already removed. Reinstall nut (10) and bump pinion from brake side toward differential until free from housing.

On all models, inspect the axle shaft (32) and pinion shaft against the values which follow:

850 Models

Axle shaft (32), OD at bearings
(21 & 24) . . . . 50.00-50.02 mm
Pinion shaft (1)-
  OD at bearing (4) . . . . 35.00-35.02 mm
  OD at inner bearing . . 29.94-29.96 mm

950 and 1050 Models

Axle shaft (32), OD at bearings
(21 & 24) . . . . 60.00-60.02 mm
Pinion shaft (1)-
  OD at bearing (4) . . . . 40.00-40.02 mm
  OD at inner bearing . . 32.96-32.98 mm

Observe the following, when assembling. Install narrow outer seal (26 – Fig. 152) with lip toward outside, then install thicker seal (25) with lip toward bearing (24). Use sealer to stick gasket (27) onto housing (28). Coat lips of seals (25 & 26) with grease. Position assembled housing and seals over axle (32) and install seal wear sleeve (31) on axle with larger inside diameter, to accept "O" ring (30), toward small end of axle. Install "O" ring (30) and retainer ring (29) in counter bore of sleeve (31), then install bearing (24) over axle. Bearing should be heated to 150 degrees C. before installing and should have retainer ring on outside diameter toward outer end of axle. Be sure that wear sleeve (31) and bearing (24) are both sealed. Coat gasket (27)

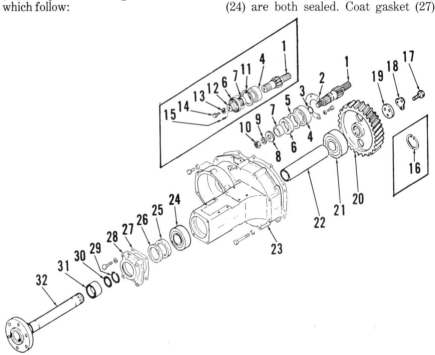

Fig. 152 — Exploded view of final drive and axle assembly. Parts shown in insets are used on 850 models.

1. Final drive pinion	10. Nut (Except 850 models)	17. Screws (Except 850 models)
2. Retainer	11. Snap ring (850 models)	18. Lock plate (Except 850 models)
3. "O" ring	12. Washer (850 models)	19. Retainer plate (Except 850 models)
4. Bearing	13. Lockwasher (850 models)	20. Final drive gear
5. Washer	14. Screw (850 models)	21. Axle inner bearing
6. Oil seal	15. Pin (850 models)	22. Spacer
7. Spacer	16. Snap ring (850 models)	23. Axle housing
8. Washer (Except 850 models)		24. Axle outer bearing
9. Washer (Except 850 models)		25. Oil seal
		26. Dust seal
		27. Gasket
		28. Retainer
		29. Snap ring
		30. "O" ring
		31. Seal sleeve
		32. Axle

with sealer and insert the axle, seals and outer bearing (24 through 32) into final drive housing (23). Tighten retainer attaching screws to 27 N·m torque. Install sleeve (22) and drive bearing (21) onto shaft. make sure that inner race of bearing (21) is seated against sleeve (22) on axle and that axle will turn freely in bearings.

On 850 models, install seal (6) with lip toward gear of pinion (1). Drive seal into bore, install snap ring (11), then drive seal in against snap ring. Press bearing (4) onto pinion shaft (1) until tight against shoulder. Coat inside and outside diameters of seal collar (7) with grease and press onto pinion shaft tightly against inner bearing race. Install pinion and bearing assembly in housing (23) until bearing outer race is against snap ring (11). Install the two screws which retain bearing (4) in housing to 15 N·m torque. Install final drive gear (20) on axle, then retain with snap ring (16). Measure clearance between snap ring (16) and hub of gear (20), then install shims equal to the thickness between inner race of bearing (21) and gear (20). Screw (14) which attaches brake drum should be tightened to 54 N·m torque.

On 950 models, install seal (6) with lip toward gear of pinion (1). Coat "O" ring (3) with oil and install in groove of pinion shaft. Lubricate shaft (1), bearing (4) and seal sleeve (7) with oil, then press parts (4 and 7) over shaft and "O" ring. Position washer (5) in bore over installed seal (6), then install pinion, bearing and wear sleeve. Be sure that outer race of bearing is seated against shoulder in housing, then install retainer (2). Tighten retainer attaching screws to 27 N·m torque. Install final drive gear (20), spacer washer (19), lock plate (18) and screws (17). Tighten screws (17) to 54 N·m, then lock by bending plate (18) around flats of screw heads. Brake drum retaining nut (10) on 950 models should be tightened to 167-245 N·m torque.

On 1050 models, two opposing seals are used at (6) on pinion shaft. Open side of outer seal faces brake drum. Open side of inner seal faces washer (5) and pinion gear bearing (4). Other assembly procedures are the same as for 950 models described in preceding paragraph. Brake drum retaining nut (10) should be tightened to 300-350 N·m torque.

BRAKES

ADJUST

All Models

154. Brake pedals should have approx-

imately 25 mm free travel and pedals should align when equal pressure is applied to the two pedals. Pedal free play should never exceed 35 mm. Adjustment is accomplished by shortening or lengthening actuating rods. Loosen locknut and turn turnbuckle located in middle of each rod.

If actuating rod can not be adjusted to limit free play to less than 35 mm, brakes can be disassembled and brake shoes rotated. The leading end of shoes will wear faster. The anchor pin (4 – Fig. 154) can be turned 90 degrees to position the brake shoes closer to the drum after loosening nut (7).

R&R AND OVERHAUL

All Models

155. To remove the brake cover (8 – Fig. 154) and related parts (2 through 10), proceed as follows. Detach actuating rod from lever (10), then unbolt and remove cover from final drive housing. Shoes (2) can be removed by removing springs (3).

If brake compartment is contaminated with oil, refer to paragraphs 150 and 151 for renewal of pinion shaft seal or seals (6 – Fig. 152). If contaminated with moisture, be sure that hose for vent (11) is correctly routed for early models or that vent tube points down on later models without hose. Refer to the following specification data and Fig. 154.

Operating Cam (10) Outside Diameter –
   850 Models . . . . 21.95-21.98 mm
   950 & 1050 Models . . . 27.95-27.98 mm
Anchor Pin (4) Outside Diameter –
   850 Models . . . . 13.97-14.00 mm
   950 & 1050 Models . . . 21.98-22.00 mm
Anchor Pin (4), All Models
   Width across flats . . . . . 16.9-17.1 mm
   Thickness at end . . . . . . 13.9-14.1 mm

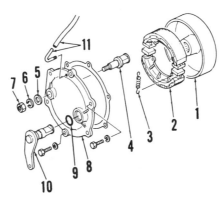

Fig. 154 – Exploded view of typical brake assembly.

1. Brake drum
2. Brake shoes
3. Return springs
4. Anchor pin
5. Washer
6. Lockwasher
7. Nut
8. Brake plate
9. "O" ring
10. Actuating cam

Brake Springs (3), Working Load-
   850 Models . . . . 249 N @ 52 mm
   950 & 1050 Models . . . 271 N @ 62 mm
Brake Lining Thickness, New –
   850 Models . . . . 4.80 mm
   950 & 1050 Models . . . . 4.75 mm
Brake Lining Thickness, Wear Limit –
   All Models . . . . 3.00 mm
Brake Drum, Inside Diameter –
   850 New . . . . 140-141 mm
   850 Wear Limit . . . . 142 mm
   950 & 1050 New . . . . 170-171 mm
   950 & 1050 Wear Limit . . . . . . 172 mm
Brake Drum Retaining Nut or Screw,
   Torque –
   850 Models (Screw) . . . . . . 49-60 N·m
   950 & 1050 Models (Nut) 168-245 N·m

When reinstalling, engage brake to center cover over drum, then tighten screws retaining cover (8) to axle housing to 23-30 N·m torque for 850 models, 49-59 N·m torque for 950 and 1050 models. Refer to paragraph 154 for complete adjustment, including setting anchor pin (4) which should be assembled at minimum thickness with new brake shoes.

POWER TAKE OFF

OUTPUT SHAFT

All Models

160. **REMOVE AND REINSTALL.** To remove pto output shaft, first remove rockshaft housing as outlined in paragraph 205, pto pinion shaft as in paragraph 161 or 162, left final drive as outlined in paragraph 150 and pto shift linkage as outlined in paragraph 165.

Remove four screws retaining oil seal housing to rear of center housing. Lift snap ring (20 – Fig. 160) from its groove and slide snap ring and shift collar (21) forward on shaft. Lift snap ring (22) from groove and move it (forward) toward bearing (19). Attach puller to output shaft (26) and pull shaft toward rear. Remove snap ring (18), just as soon as bearing has moved to rear enough to remove snap ring. Remove parts as shaft is pulled to rear.

Inspect all parts for wear or obvious damage. Inside diameter of gear (24) should be 45.07-45.09 mm, outside diameter of bushing (25) should be 44.99-45.00 mm, inside diameter of bushing (25) should be 39.99-40.00 mm.

Install thrust washer (23R) with grooves on washer toward front (gear 24), then press bearing (27) onto shaft, tight against washer (23R). The snap ring on bearing (27) must be toward rear. Install spacer washer (28) and snap ring (29). Install wear ring (30) tight against snap ring (29) and snap ring (31) in groove behind wear ring. Insert shaft through rear opening, then position gear (24), thrust washer (23), snap ring (22), shift collar (21) and snap ring (20) over shaft. Spur cut inside diameter of gear (24) should be toward front (collar 21) and groove of shift collar (21) should be forward. Install bearing (19) on end of shaft and snap ring (18) in groove. A sheet of steel plate can be used to cover hole in case for bearing (19) to hold bearing while bumping shaft from rear into bearing. A 1⅛ inch open end wrench can be used to hold bearing (19) while clearing shaft to finish installation of bearing. Bearing should bottom on shoulder of shaft before installing snap ring (18). Bump shaft forward until snap ring for bearing (27) is seated against housing and bearing (19) is in bore at front. Move gear (24) and thrust washer (23) to rear and install snap ring (22) in proper groove in shaft. Slide shaft collar (21) toward rear and engage coupling splines with gear (24). Install snap ring (20) in second groove of shaft splines. Spring loaded lip of seal (33) should be toward inside (bearing 27). Closed side of seal (33) should be against shoulder of bore in retainer (34). Install "O" ring (32) in groove. Screws attaching retainer (34) to transmission housing should be tightened to 27 N·m torque. Assembly is completed by installing shifter assembly (36 through 46), pinion shaft (1 through 9) and rockshaft housing.

DRIVE PINION SHAFT

850 and 950 Models

161. **REMOVE AND REINSTALL.** To remove the pto drive pinion shaft (4 – Fig. 160) and related parts, first remove rockshaft housing as outlined in paragraph 205. Unbolt and remove the pto shaft shield from rear of transmission/differential housing, then unbolt and remove retainer (1). Attach a puller (such as John Deere special tool JDE-114-1) to threaded bore in rear end of shaft (4) and pull shaft out toward rear. Remove parts (5, 6, 7, 8 & 9) as shaft is pulled to rear and be careful to remove any parts that are dropped.

Inside diameter of bushing (12) should be 18.00-18.018 mm and outside diameter should be 22.435-22.456 mm. Bushing (12) should have oil hole aligned and should be flush with bottom of bore chamfer. Free length of spring (7) is 74 mm and should exert 66 N force when compressed to 29 mm. Snap ring (15R) is not used on late models or replacement shafts (4).

Hold pin (5) in hole of shaft (4) with grease while assembling. Pin (5) aligns with notch in washer (9). Assemble bear-

ing (3) and snap ring (15R) on shaft. Insert shaft through rear opening, while positioning washer (6), spring (7) and coupling half (8) over shaft. Be sure that pin (5) is still correctly positioned, then install thrust washer (9) over pin just before shaft (4) enters bushing (12). Grooved side of washer (9) should be toward front (coupling 11). Coat gasket (2) with sealer and install cover (1). Retaining screws for cover (1) should be tightened to 27 N·m torque. Be sure that washer (6) compresses spring (7) against coupling (8). Snap ring (15R), if used, should be in front groove of early shafts (4). Late shafts do not use snap ring (15R), but use shoulder instead.

1050 Models

162. **REMOVE AND REINSTALL.** To remove the pto drive pinion shaft (413 – Fig. 160) and bearings (3 & 14),

first remove rockshaft housing as outlined in paragraph 205. Unbolt and remove pto shaft shield from rear of transmission/differential housing, then unbolt and remove retainer (1). Attach a puller (such as John Deere special tool JDE-114-1) to threaded bore in rear end of shaft (413) and pull shaft out toward rear.

Reinstall, by positioning front bearing (14) in case bore, then installing shaft (413) and rear bearing (3). Be sure that splines on shaft (413) correctly engage coupling (17). Install gasket (2) and cover (1). Tighten retaining screws to 27 N·m torque.

PTO DRIVE SHAFT

850 and 950 Models

163. **REMOVE AND REINSTALL.** The drive shaft (13 – Fig. 160) can be re-

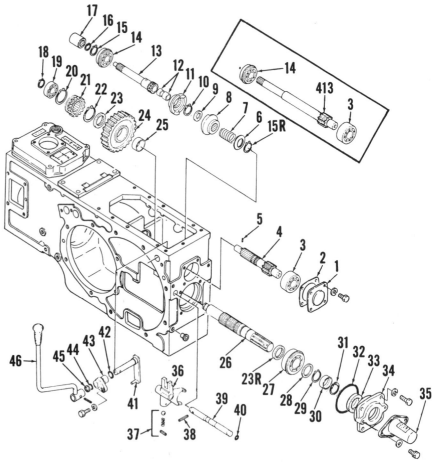

Fig. 160 – Exploded view of 850 and 950 pto drive. Inset shows parts used on 1050 models in place of parts (4 through 17).

1. Cover	13. Shaft	24. Gear	36. Shift fork
2. Gasket	14. Bearing	25. Bushing	37. Detent
3. Bearing	15. Snap ring	26. Output shaft	38. Pin
4. Shaft	16. Snap ring	27. Bearing	39. Shaft
5. Pin (3.5 x 6 mm)	17. Coupling	28. Washer	40. "O" ring
6. Washer	18. Snap ring	29. Snap ring	41. Shift lever
7. Spring	19. Bearing	30. Seal sleeve	42. "O" ring
8. Clutch	20. Snap ring	31. Snap ring	43. Bearing
9. Washer	21. Sliding gear	32. "O" ring	44. "O" ring
10. Snap ring	22. Snap ring	33. Seal	45. Back-up washer
11. Clutch	23. Thrust washer	34. Retainer	46. Lever
12. Bushings	23R. Thrust washer	35. Cover	413. Shaft (1050 models)

moved after first removing drive pinion shaft as outlined in paragraph 161. If transmission is separated from clutch housing, a long brass drift can be inserted through transmission to drive shaft (13) and bearing (14) out toward rear. If clutch and transmission housings are not separated, drive bearing (14) toward rear or use claw type puller around coupling (11). Be careful not to let coupling (17) fall when shaft is removed.

Inside diameter of bushing (12) should be 18.00-18.018 mm and outside diameter should be 22.435-22.456 mm. Bushing (12) should have oil hole aligned and should be flush with bottom of hole chamfer.

When assembling, position coupling (17) on rear of clutch shaft, then bump shaft (13) and bearing (14) into position. Snap ring on bearing (14) should be toward rear and should be seated against housing. Finish assembly by installing pto drive pinion (14) and associated parts as described in paragraph 161.

SHIFT LINKAGE

All Models

165. REMOVE AND REINSTALL. The shift linkage (36 through 46–Fig. 160) can be removed after removing rockshaft housing as in paragraph 205 and the left final drive assembly as outlined in paragraph 150. Drive roll pin from external lever (46), then pull shaft (41) toward center of housing and remove. Drive roll pin (38) from housing and shaft (39), then push shaft (39) toward rear out of housing. Remove shift fork (36) and detent assembly (37) as shaft (39) is withdrawn. Be careful to catch the detent ball as shaft is withdrawn from fork.

Free length of detent spring is 33.8 mm and spring should exert 72-79 N force when compressed to 26.4 mm. Shift fork thickness should be 6.7-6.9 mm and groove in shift collar (21) should be 7.1-7.3 mm. Install new fork and/or shift collar if clearance exceeds 1 mm. Outside diameter of shaft (39) is 14.96-14.98 mm and inside diameter of bore in fork (36) should be 15.00-15.04 mm. Renew fork and/or shaft if clearance exceeds 0.2 mm.

When assembling, roll pin for detent (37) should be installed with split toward ball. A short rod approximately 14.95 mm diameter can be used to hold detent compressed while inserting rod (39) through fork (36). Be sure that assembly rod is short enough to be removed. Split in pin (38) should be toward front and pin should be 3mm below flush with outside of housing.

Lubricate "O" ring (42) and sleeve (43), position "O" ring in groove of sleeve, then install assembly in housing bore. Insert arm shaft (41) through sleeve, engaging arm with notch in shift fork. Install "O" ring (44) and washer (45), then install lever (46). Attaching screw for sleeve (43) should be tightened to 27 N·m torque.

HYDRAULIC LIFT SYSTEM

FLUID AND FILTER

All Models

170. The transmission, differential and hydraulic system share a common sump. Lubricant should be maintained at full level marked on dipstick. The dipstick is attached to fill plug located in transmission shift cover. Do not screw plug in when checking fluid level. Oil should be drained, strainer (3–Fig. 170) should be cleaned and system should be filled with new fluid every 200 hours of operation. Every 600 hours of operation, strainer (3) and filter (9) should be renewed. Capacity is 18 liters for 850 and 950 models, 26 liters for 1050 models. Allow sufficient time for oil to pass between compartments and be sure that tractor is level before checking fluid level with dipstick. Use only John Deere Hy-Gard Transmission and Hydraulic Oil or equivalent.

HYDRAULIC PUMP

All Models

171. REMOVE AND REINSTALL. To remove the hydraulic pump, first clean the pump, fittings and surrounding area to prevent dirt from entering the system. Unbolt inlet and outlet fittings from the pump, then unbolt pump from engine front cover.

Pump drive coupling can be removed from pump shaft after removing the retaining nut.

Repair parts are not available for hydraulic pump and disassembly is not recommended.

When reassembling, install pump drive coupling, flat washer, lockwasher, then retaining nut. Tighten coupling retaining nut to 15 N·m torque.

Install pump, with arrow on pump front cover at eight o'clock position. Tighten pump mounting screws to 26 N·m torque and pipe attaching screws to 10 N·m torque.

Pressure tests require a gage of sufficient capacity to withstand higher than correct pressure. Pressure gage with capacity of 34474 kPa (5000 psi) is recommended for testing.

TESTS AND ADJUSTMENTS

All Models

172. A complete systematic check is necessary for tracing many hydraulic system problems; however, not all tractors are equipped with all possible components. Before testing pressure and flow, visually check system hoses, tubes, fittings and components for evidence of leakage or other damage. Check fluid level and condition of fluid, then add or change fluid and service filters if necessary before continuing. Start the engine and operate at approximately 1500 to 1700 rpm. Turn stop valve (17–Fig. 175 or Fig. 181) in and move rockshaft control lever to rear so that oil will be forced through relief valve. Continue to operate until hydraulic fluid reaches 38-49 degrees C. (100-120 degrees F.) before beginning pressure and flow tests.

Fig. 170–Exploded view of inlet filter screen (1 through 8) and return oil filter (9 through 21) used on all models.

1. Cover
2. Gasket
3. Filter screen
4. Drain plug
5. Housing
6. "O" rings
7. Side cover
8. Plug
9. Filter element
10. Retainer
11. Bearing
12. Washer
13. "O" ring
14. Spring
15. Housing
16. Hose connector
17. Elbow
18. Bracket
19. "O" ring
20. Connector
21. Hose

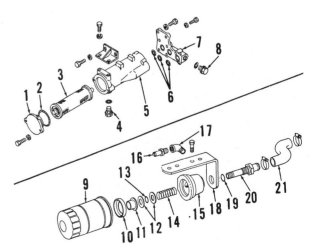

SYSTEM TESTS

850 and 950 Models

173. **POWER STEERING.** Refer to paragraph 15 to check and adjust power steering relief valve of models so equipped.

174. **SELECTIVE CONTROL VALVE RELIEF VALVE.** Attach a pressure gage to diverter block as shown in Fig. 173. Run engine at 1500 rpm, move the selective control valve lever forward or rearward and check relief valve setting. Pressure should be 16892-17582 kPa (2450-2550 psi). If incorrect, add or remove shims (2 – Fig. 174) as necessary to establish correct pressure. If pressure is too low and can not be increased by adding shims to relief valve, filter screen (3 – Fig. 170) may be plugged or pump may be damaged.

175. **SYSTEM FLOW TEST.** Attach flow meter to elbow on top of diverter valve and insert outlet hose from flow meter into fill opening in transmission shifter cover. It is recommended that hose be tied in such a way to hold hose in fill opening so that hose will not accidentally come out of fill opening when testing. Operate engine at 2400 rpm, adjust control valve on flow meter to obtain test pressure of 10342 kPa (1500 psi), then observe rate of flow. Correct flow is more than 18.9 liters per minute (5 gpm). If flow is too low, check pump inlet screen (3 – Fig. 170) for plugging. If filter screen is not plugged, check for leaking relief valve (5 – Fig. 174), scored selective control valve (17 or 30) or cracked housing (32).

176. **ROCKSHAFT RELIEF VALVE.** Attach pressure gage to test port on diverter valve as shown in Fig. 173 on models with power steering or selective control valve. On models without selective control valve or power steering, attach pressure gage to port

normally covered by plug (2 – Fig. 175). On all models, insert a screwdriver or similar tool between feedback lever (46 – Fig. 176) and control valve rod (57). Turn stop valve (17 – Fig. 175) completely out, start engine, move rockshaft control lever rearward, set engine speed to 1500 rpm, then observe pressure on gage. Pressure should be 12755-13445 kPa (1850-1950 psi) with fluid temperature at 38-49 degrees C. (100-120 degrees F.).

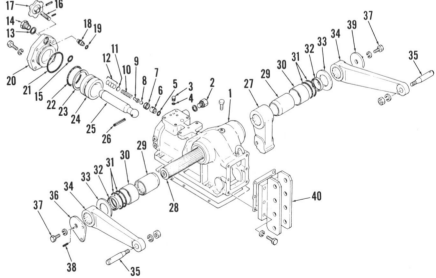

Fig. 174 – Exploded view of remote control valve used on 850 and 950 models. Casting marks "IN", "OUT" and "BYD" are located as shown near ports.

1. Plug
2. Shim
3. "O" ring
4. Spring
5. Relief valve poppet
6. "O" ring
7. Connector fitting
8. Plunger
9. Spring
10. "O" ring
11. Plug
12. Connector
13. "O" ring
14. Back-up ring
15. "O" ring
16. "O" rings
17. Spool
18. Bushing (same as 31)
19. Shim
20. Shim
21. Spring
22. Spacer
23. Washer
24. Retaining clip
25. Cap
26. Cover
27. Cap
28. Balls (2 used – 3.97 mm)
29. Spring
30. Valve spool
31. Bushing (same as 18)
32. Housing

Fig. 173 – Refer to text for checking hydraulic pressures.

Fig. 175 – Exploded view of 850 and 950 rockshaft, housing and associated parts.

1. Rockshaft housing	11. Guide	21. Gasket	31. "O" rings
2. Plug	12. Shims	22. Piston ring	32. Seals
3. Plug	13. Seal washer	23. Back-up ring	33. Washers
4. "O" ring	14. Plug	24. Piston	34. Lift arms
5. "O" ring	15. "O" ring	25. Rod	35. Shafts
6. Valve seat	16. Shaft	26. Roll pin	36. Plate
7. Guide	17. Stop valve	27. Crank arm	37. Retaining screw
8. Ball	18. Connector	28. Rockshaft	38. Roll pin
9. Holder	19. "O" ring	29. Splined spacers	39. Washer
10. Spring	20. Cylinder cover	30. Bushings	40. Bracket

If pressure is too high, remove plug (14 – Fig. 175) and reduce thickness of shims (12). Recheck pressure after removing shims. Each shim should change pressure approximately 1193 kPa (28 psi).

If pressure is too low, check for leakage as follows: Turn stop valve (17) in, operate engine at 1500 rpm and observe gage pressure. If pressure is still less than 12755 kPa (1850 psi), remove cylinder cover and piston. Check cylinder bore, "O" ring (22), back-up ring (23) and piston (24) for damage. If cylinder is not leaking, shims (12), may be added to increase pressure to limits of 12755-13445 kPa (1850-1950 psi).

If pressure can not be adjusted, check "O" rings (13 & 15 – Fig. 174) and back-up ring (14). If "O" rings appear satisfactory, remove rockshaft control valve (63 through 69 – Fig. 176) and inspect for scoring. If rockshaft valve is not damaged, remove control valve housing and check condition of "O" ring (72), plug (3) and "O" ring (4). If "O" rings (72 & 4) are not damaged, remove cylinder cover (20 – Fig. 175) and check condition of ball (8), seat (6) and "O" ring (5). If pressure still can not be adjusted and ball (8), seat (6) and "O" ring (5) are in good condition, install new hydraulic pump and recheck.

Be sure to remove screwdriver which was inserted between feedback lever (46 – Fig. 176) and control valve rod (57) after test is completed.

177. ROCKSHAFT LIFT CYCLE. Move rockshaft control lever forward, and turn stop valve (17 – Fig. 175) out completely. Attach 226 kg (550 lbs.) of weight to rockshaft lift arms, run engine at 2400 rpm, move rockshaft control lever toward rear and record time required to move lift arms from fully lowered position to full raised position. Lift arms should raise fully within 2.5-3.0 seconds with hydraulic fluid at 38-49 degrees C. (100-120 degrees F.).

If rockshaft chatters, check condition of inlet filter screen (3 – Fig. 170). If inlet filter screen is not plugged, check condition of "O" ring (72 – Fig. 176), plug (3) and "O" ring (4). If "O" rings (72 & 4) are not damaged, check condition of ball (8 – Fig. 175), seat (6) and "O" ring (5). If system still will not raise lift arms correctly, pump may be faulty.

If lift arms do not raise in 2.5-3.0 seconds, remove rockshaft control valve (63 through 69 – Fig. 176) and inspect for scoring. If rockshaft valve is not damaged, remove control valve housing and check condition of "O" ring (72), plug (3) and "O" ring (4) as described in the preceding paragraph.

System can be checked for flow as out-

lined in paragraph 175 to further check for damaged pump.

1050 Models

180. POWER STEERING. Refer to paragraph 15 to check and adjust power

steering relief valve of models so equipped.

181. SYSTEM RELIEF VALVE TEST. Turn stop valve (17 – Fig. 181) counter-clockwise as far as possible, attach pressure gage to port in-place-of plug (2). Run engine at 1500 rpm, hold

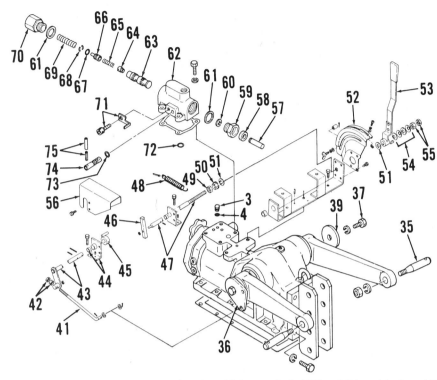

Fig. 176 — Exploded view of rockshaft controls and linkage typical of 850 and 950 models. Parts (47 and 49 through 55) are different on some models.

3. Plug	45. Collar	56. Cover	67. "O" ring
4. "O" ring	46. Arm	57. Push rod	68. Retaining clips
35. Shafts	47. Control shaft	58. Seal	69. Spring
36. Plate	48. Spring	59. Connector	70. Plug
37. Retaining screw	49. Locknut	60. Snap ring	71. Slow return valve
39. Washer	50. Nut	61. Seal (2 used)	retainer
41. Rod	51. Washer	62. Valve housing	72. "O" ring
42. Adjusting and	52. Control lever guide	63. Valve spool	73. "O" ring
locknuts	53. Control lever	64. Poppet	74. Slow return valve
43. Control arm and shaft	54. Spring washers	65. Spring	75. Handle (pins)
44. Bushing	55. Nuts	66. Retainer	

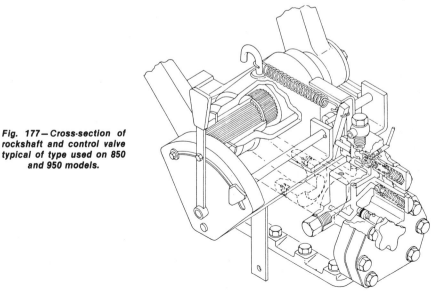

Fig. 177 — Cross-section of rockshaft and control valve typical of type used on 850 and 950 models.

the load and depth sensing lever (41) down in "LOAD" position, hold the rockshaft control lever (45) in "RAISE" position and observe gage pressure. Pressure should be 14134-14325 kPa (2050-2150 psi) with oil temperature 38-49 degrees C. (100-120 degrees F.).

If pressure is too high, remove one shim (19 – Fig. 182) and recheck pressure.

If pressure is too low, lower rockshaft, close stop valve (17 – Fig. 181) by turning clockwise as far as possible, move control lever (45) rearward to "RAISE"

position and recheck pressure. If pressure is still too low, add one shim (19 – Fig. 182) and recheck. If pressure is still too low recheck condition of parts (20 through 24). If parts of relief valve are not damaged, check system flow as described in paragraph 182. If pressure was normal when checked with stop valve (17 – Fig. 181) closed, remove cylinder cover (20) and check cylinder walls for scoring or other damage. If cylinder walls are not scored, check surge relief valve (5, 6, 7, 9, 10, 11 and 12) for damage. If surge relief valve

does not appear to be damaged, or is repaired, surge relief valve should be adjusted as outlined in paragraph 183 before reassembling. If surge relief valve needed repair or adjustment, reassemble and recheck pressures. If surge relief valve tests indicate valve is satisfactory, remove rockshaft piston, install new back-up ring (23) and "O" ring (22), then reassemble and recheck pressure.

182. **SYSTEM FLOW TEST.** On models with only rockshaft hydraulics, remove plug from lower front of hydraulic junction block and install a ¼-inch NPT Allen plug in small inner hole. Install John Deere fitting (part number 6749) or equivalent in lower plug hole and attach flow meter inlet to this fitting. Route outlet hose from flow meter to the filler opening in transmission shift cover.

On models with either selective control valve or power steering or both, disconnect tube (R – Fig. 183) from lower front of hydraulic junction block. Use necessary fittings to attach flow meter to port in lower front of junction block. Route outlet hose from flow meter to

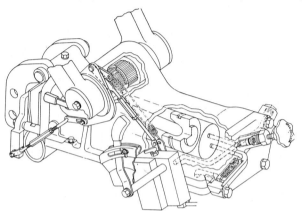

Fig. 180 — Cross-section of rockshaft and controls typical of type used on 1050 models.

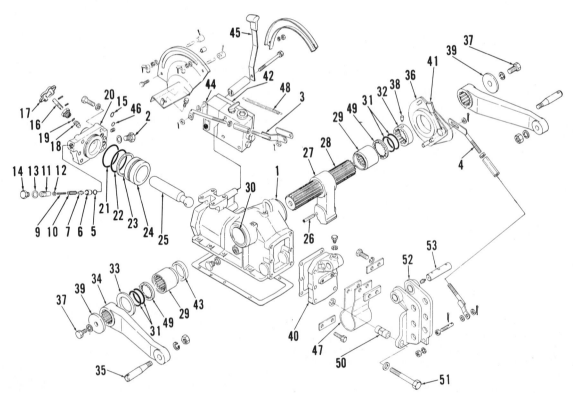

Fig. 181 — Exploded view of rockshaft and related parts typical of 1050 models.

1. Rockshaft housing	19. "O" ring	28. Rockshaft	45. Control lever
2. Plug	20. Cylinder cover	29. Splined spacers	46. "O" ring
3. Link	21. Gasket	30. Bushings	47. Spring
4. Link	22. Piston ring	31. "O" rings	48. Spring
5. "O" ring	23. Back-up ring	32. Bushing	49. Seals
6. Valve seat	24. Piston	33. Washer	50. Nuts
7. Guide	25. Rod	34. Lift arm	51. Screws
9. Pin	26. Roll pin	35. Shafts	52. Bracket
10. Spring	27. Crank arm	36. Sensing plate	53. Pivot
11. Guide		37. Retaining screw	
12. Shims		38. Pin	
13. "O" ring		39. Washers	
14. Plug		40. Bracket	
15. "O" ring		41. Sensing lever	
16. Shaft		42. Control valve	
17. Stop valve		43. Spacer	
18. Connector		44. Lever	

filler opening in transmission shift cover.

On all models, open flow meter shut-off valve, operate engine at 2400 rpm, then adjust flow meter shut-off valve to set pressure at 10342 kPa (1500 psi). New pump should provide approximately 18.9 liters per minute (5.0 gpm) with fluid at 38-49 degrees C. (100-120 degrees F.). If flow is less than 15.1 liters per minute (4.0 gpm), remove pump inlet screen (3–Fig. 170) and check for plugging. If inlet screen is not plugged, install new pump and recheck flow.

Remove flow meter from all models after testing. On models with only rockshaft hydraulics, be sure to remove previously installed ¼-inch Allen plug, before installing larger plug in lower front port of junction block.

183. ROCKSHAFT RELIEF VALVE. To check surge relief pressure, remove cylinder cover (20–Fig. 181), remove relief valve (5, 6, 7, 9, 10, 11, 12, 13 and 14) and check parts especially valve (7), seat (6) and "O" ring (5) for damage. Tap hole inside cover for ⅛-inch NPT, clean debris, then assemble surge relief valve using previously installed shims (12). Attach a suitable hand pump to the tapped hole (⅛-inch NPT) in center of cover as shown in Fig. 184. Operate hand pump and observe pressure relief setting. Surge relief pressure should be 17237-20684 kPa (2500-3000 psi). If pressure is too low, add one shim (12), then recheck. If pressure is too high, remove one shim (12) and recheck.

184. ROCKSHAFT LIFT CYCLE TEST. Move rockshaft control lever forward and turn stop valve (17–Fig. 181) completely out (counter-clockwise). Attach 226 kg (500 lbs.) of weight to the rockshaft lift arms, operate engine at 2400 rpm. Move the rockshaft control lever fully to the rear and record the time required to raise the lift arms. On models **without** power steering, lift arms should move from full lowered to full raised in 2.5-3.0 seconds. On models **with** power steering, full travel from lowered to raised should take lift arms 3.0-3.5 seconds.

If rockshaft chatters or if lift is too slow, check flow as outlined in paragraph 182.

TROUBLE SHOOTING

All Models

190. Refer to the following when diagnosing hydraulic system troubles.

Rockshaft Cannot Be Lifted or Lift Speed Is Slow.
Relief pressure too low (paragraph 174 or 181).
"O" ring (5–Fig. 175 or Fig. 181) broken.
"O" ring (22–Fig. 175 or Fig. 181) or back-up ring (23) damaged.
Clogged inlet filter screen (3–Fig. 170).
Hydraulic fluid incorrect.
Hydraulic pump worn (paragraph 171).
Return spring (48–Fig. 176 or Fig. 181) too weak.
Crank arm (27–Fig. 175 or Fig. 181) broken.

Rockshaft Cannot Be Lowered
Return spring (48–Fig. 176 or Fig. 181) broken or detached.
Slow return valve (74–Fig. 176 or 1–Fig. 182) stuck closed.
Debris in spool valve (63–Fig. 176 or 32 and 41–Fig. 182).

Neutral Position Unstable, Rockshaft Falls When Engine is Turned Off
Broken "O" ring (19 or 22–Fig. 175 or Fig. 181).
Spool valve (63–Fig. 176 or 32 and 41–Fig. 182) worn.
Return spring (48–Fig. 176 or Fig. 181) too weak.

Control Valve Noisy
Feedback rod (41–Fig. 176 or 3 and 4–Fig. 181) improperly adjusted.
Improper hydraulic fluid used.

LINKAGE ADJUSTMENTS

850 and 950 Models

200. Be sure that spring (48–Fig. 176) has correct tension of 271 N. at 190.5 mm (61.3 lbs. at 7.5 inches). Tighten friction adjusting nuts (55) until 31-58 N. (7-13 lbs.) force is required to move lever (53). Adjust length of rod (41) to obtain 7-10 mm (0.28-0.40 inch) play measured at end of lift arms.

1050 Models

202. Distance (A–Fig. 202) should be adjusted by turning both screws (51–Fig. 181) until distance is exactly 131 mm, then tighten nuts (50). This adjustment will set tension of load sensing spring (47).

Attach a spring scale 160 mm from pivot of lever (45), then check force required to move lever. Lever should move with 3.2-8.2 kg (7-18 lbs.) force. Friction is changed by adding or removing spring washers (60–Fig. 182).

Length of rod (3–Fig. 181 and Fig.

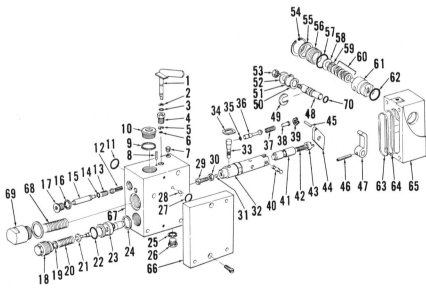

Fig. 182 – Exploded view of control valve used on 1050 models.

1. Slow return valve	19. Shim	36. Lowering valve	53. Nut
2. Wave washer	20. Spring	37. Spring	54. Snap ring
3. Shim	21. Relief valve	38. Stop pin	55. Washer
4. Bushing	22. "O" ring	39. Snap ring	56. Cap
5. Snap ring	23. Relief valve spool	40. Pin	57. "O" ring
6. "O" ring	24. Gasket	41. Shut-off spool	58. Washer
7. Plug	25. Gasket	42. Spring	59. Shim
8. Roll pin	26. Plug	43. Spring seat	60. Springs
9. Gasket	27. "O" ring	44. Positioning plate	61. Control shaft
10. Plug	28. Plug	45. Pivot pin	62. "O" ring
11. "O" ring	29. Screw	46. Spring pin	63. Leaf spring
12. Check valve	30. Locknut	47. Driver lever	64. "O" ring
13. Spring	31. Plug	48. Driver shaft	65. Selector housing
14. "O" ring	32. Pilot spool	49. Snap ring	66. Cover
15. Valve stop	33. Eccentric pin	50. Washer	67. Valve housing
16. Gasket	34. Washer	51. Snap ring	68. Spring
17. Plug	35. "O" ring	52. Lockwasher	69. Plug
18. Plug			

202) should be 260 mm as measured between rod ends. Length of rod (4) should be 211 mm as measured from rear of rod slot to rear attaching point.

Move sensing lever (41–Fig. 181) to rear, then move control lever (45) to rear to raise lift arms. Adjust length of rod (4) to provide approximately 12-25 mm free play at top of travel. Distance (B–Fig. 202) from lift arm holes to gasket surface of rockshaft housing should be approximately 382 mm.

ROCKSHAFT HOUSING

All Models

205. **REMOVE AND REINSTALL.** To remove the rockshaft housing from top of transmission/differential housing, proceed as follows:

Lower rockshaft and remove hydraulic line from top of control valve. Remove lift links and tractor seat. Unbolt rockshaft from transmission/differential cover, then lift unit away from tractor.

When reinstalling, tighten rockshaft housing to transmission/differential housing screws to 57 N·m torque.

RELIEF VALVE, PISTON COVER AND PISTON

850 and 950 Models

206. **REMOVE AND REINSTALL.** The relief valve is located at (5 through 14–Fig. 175). Shims (12) used to adjust relief pressure can be added or removed after removing plug (14). Removal of seat (6), "O" ring (5) and guide (7) necessitates removal of piston cover (20).

Lower rockshaft, remove plug (14), shims (12), spring (10) and holder (9). Unbolt and remove cover (20), then pull stop valve connector (18) and "O" ring (19). Rockshaft piston (24) can be removed by rotating lift arms (34) downward quickly to propel piston out of cylinder. Remove valve seat guide (7) and withdraw seat (6), "O" ring (5) and ball (8).

Check seat (6), holder (9) and ball (8) for damage. Spring (10) should exert 369 N when compressed to 50 mm. Inspect piston (24) and cylinder bore for evidence of scoring or other damage. Outside diameter of piston is 79.94-79.97 mm and inside diameter of cylinder is 80.00-80.05 mm. Back-up ring (23) and "O" ring should be installed in piston groove in order shown. Outside diameter of stop connector (18) is 11.93-12.00 mm and inside diameter of bore for connector is 12.00-12.03 mm.

Lubricate all parts as they are being installed. Install piston and cylinder cover. Tighten cylinder cover retaining screws to 95 N·m torque and relief valve plug (14) to 97-127 N·m torque. Be sure that copper washer (13) is new, before tightening.

NOTE: It is important to open stop valve (17) when storing tractor. Do not store tractor with valve closed.

1050 Models

207. **REMOVE AND REINSTALL.** The relief valve (5, 6, 7, 9, 10, 11, 12, 13 and 14–Fig. 181) is located in piston cover (20). Shims (12) are used to adjust pressure as outlined in paragraph 183.

Lower rockshaft, then unbolt and remove cover (20). Rockshaft piston (24) can be removed by rotating lift arm (34) downward quickly to propel piston out of cylinder. Relief valve can be removed, checked and adjusted as outlined in paragraph 183. Seat (6) can be removed using metal screw or similar method of pulling; however, seat should not be reused if removed in this manner.

Spring (10) free length is 42.4 mm and should exert 125 N force when compressed to 38 mm. Diameters of connector (18) are 9.9-10.1 mm and 11.93-12.00 mm. Outside diameter of piston (24) is 79.94-79.97 mm and inside diameter of cylinder bore is 80.00-80.05 mm.

Be sure that "O' ring (22) is toward closed end of piston groove and back-up ring (23) is toward open end. Tighten screws retaining cylinder cover (20) to 95 N·m torque.

ROCKSHAFT AND BUSHINGS

All Models

210. The rockshaft crank arm (27–Fig. 175 or Fig. 181) and associated parts can be disassembled after removing rockshaft housing as outlined in paragraph 205. Remove lift arms (34).

Inside diameter of bushings (30) should be 60.07-60.14 mm for 850 and 950 models, 60.01-60.02 mm for 1050 models. Splined sleeves (29) should be 59.97-59.99 mm for all models. Bushings (30) should be 8.0-8.5 mm below outer surface of housing.

Align punch marks on crank arm (27) and rockshaft (28) as these two parts are assembled in housing (1). Align punch marks on lift arms (34) and ends of rockshaft when installing lift arms. Screws (37) which attach lift arms to rockshaft should be tightened to 78-97 N·m torque. Refer to paragraph 200 or 202 for adjustment of linkage and spring (47–Fig. 181).

CONTROL VALVE

1050 Models

215. The control valve is shown exploded in Fig. 182. Disassembly will depend upon extent of repair.

Spring pin (46) should protrude 11.1 mm toward valve side of arm (47). Spring washers (60) determine friction of lever (45–Fig. 181). If too much force is necessary to move lever (45), remove one spring washer (60–Fig. 182). If lever will not stay in position, add a spring washer (60). Friction necessary to move shaft (61) should be 5-13 N·m as measured with torque wrench on bolt in center threaded hole. Use grease to hold "O" ring (64) in place while assembling. Leaf spring (63) is located inside "O" ring.

Parts (32, 41 and 67) are available only as a unit. Spring (13) should have free length of 31 mm and should exert 62 N when compressed to 13.5 mm. Open end of pilot spool (41) should be toward open end of control valve (32). Spring (42) should have free length of 47.4 mm and should exert 19.5 N when compressed to 20 mm. Eccentric pin (33) must be adjusted before installing guide (43) and

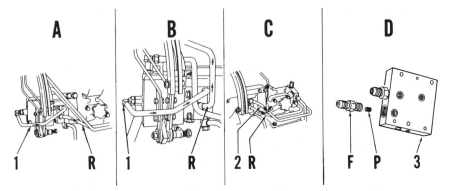

Fig. 183 – View of hydraulic junction block used on 1050 models. View "A" is for tractors with rockshaft, selective control valve and power steering. View "B" is for tractors with rockshaft and selective control valve only. View "C" is for tractors with rockshaft and power steering only. View "D" is for tractors with only rockshaft hydraulics. Tube marked "R" should be removed so that flow meter can be attached to port in junction block. Tube (1) is from "BYD" port to selective control valve. Power steering tube is shown at (2), junction block at (3), John Deere 6749 fitting at (F) and ¼ inch NPT plug, used only for testing, is shown at (P).

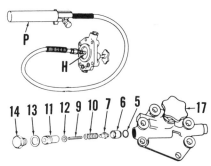

Fig. 184 — Surge relief valve can be checked and adjusted on 1050 models as described in text. Hole (H) in center of cover should be tapped so that pump (P) can be used to check relief pressure. Refer to Fig. 181 for legend.

pin (40).

Free length of relief valve spring (20)

should be 48.2 mm and should exert 492 N force when compressed to 36.5 mm.

Free length of control valve spring (68) should be 103 mm and spring should exert 251 N when compressed to 40 mm.

Free length of lowering valve spring (37) should be 39 mm and spring should exert 76 N force when compressed to 19 mm.

Assemble control valve parts (29 through 32) in housing (67), install guide washer (34), then install lowering valve parts (35 through 39). Hold special adjusting tool (John Deere part number JDG-50) against front of housing (67), then install eccentric pin (33). Turn eccentric pin until head just touches lowering valve (36), then install small lock plug (31) and tighten lock screw (29) to lock position of eccentric pin (33). Tighten locknut (30). Remainder of

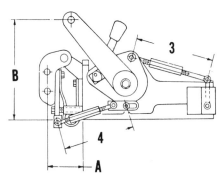

Fig. 202 — View of linkage on 1050 models showing setting dimensions. Refer to text for settings.

assembly will be self evident. Be sure to install rubber plug (28) flush with machined surface.

NOTES

NOTES

NOTES

Technical Information

Technical information is available from John Deere. Some of this information is available in electronic as well as printed form. Order from your John Deere dealer or call **1-800-522-7448**. Please have available the model number, serial number, and name of the product.

Available information includes:

- PARTS CATALOGS list service parts available for your machine with exploded view illustrations to help you identify the correct parts. It is also useful in assembling and disassembling.
- OPERATOR'S MANUALS providing safety, operating, maintenance, and service information. These manuals and safety signs on your machine may also be available in other languages.
- OPERATOR'S VIDEO TAPES showing highlights of safety, operating, maintenance, and service information. These tapes may be available in multiple languages and formats.
- TECHNICAL MANUALS outlining service information for your machine. Included are specifications, illustrated assembly and disassembly procedures, hydraulic oil flow diagrams, and wiring diagrams. Some products have separate manuals for repair and diagnostic information. Some components, such as engines, are available in separate component technical manuals
- FUNDAMENTAL MANUALS detailing basic information regardless of manufacturer:
  - Agricultural Primer series covers technology in farming and ranching, featuring subjects like computers, the Internet, and precision farming.
  - Farm Business Management series examines "real-world" problems and offers practical solutions in the areas of marketing, financing, equipment selection, and compliance.
  - Fundamentals of Services manuals show you how to repair and maintain off-road equipment.
  - Fundamentals of Machine Operation manuals explain machine capacities and adjustments, how to improve machine performance, and how to eliminate unnecessary field operations.

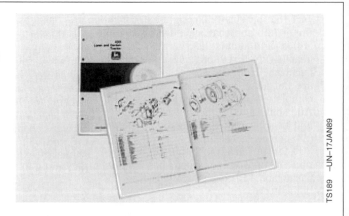

TS189 –UN–17JAN89

TS191 –UN–02DEC88

TS224 –UN–17JAN89

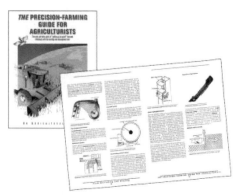

TS1663 –UN–10OCT97

DX,SERVLIT –19–11NOV97–1/1